Maíra Matos Normande

Built and Natural Heritage in a Relationship of Harmony

AF376956

Maíra Matos Normande

Built and Natural Heritage in a Relationship of Harmony

Guidelines for the Landscape Rehabilitation of Floriano Peixoto Avenue in the Historic Listed Center of the City of Penedo

ScienciaScripts

Imprint
Any brand names and product names mentioned in this book are subject to trademark, brand or patent protection and are trademarks or registered trademarks of their respective holders. The use of brand names, product names, common names, trade names, product descriptions etc. even without a particular marking in this work is in no way to be construed to mean that such names may be regarded as unrestricted in respect of trademark and brand protection legislation and could thus be used by anyone.

Cover image: www.ingimage.com

This book is a translation from the original published under ISBN 978-3-330-75606-9.

Publisher:
Sciencia Scripts
is a trademark of
Dodo Books Indian Ocean Ltd. and OmniScriptum S.R.L publishing group

120 High Road, East Finchley, London, N2 9ED, United Kingdom
Str. Armeneasca 28/1, office 1, Chisinau MD-2012, Republic of Moldova, Europe
Printed at: see last page
ISBN: 978-620-8-27544-0

Copyright © Maíra Matos Normande
Copyright © 2024 Dodo Books Indian Ocean Ltd. and OmniScriptum S.R.L publishing group

CONTENTS

SUMMARY

The diversity of architectural styles and the typological richness of the buildings on Avenida Floriano Peixoto, which have marked different periods in the urban evolution of the city of Penedo, show their importance in the historical context of the place, forming part of the Historic Center's listed universe. The singular value, both historical and artistic, of its buildings and of the street itself, demonstrates the importance of recovering this landscape, which is currently so de-characterized by the disorderly occupation of urban equipment and commercial activity and services. At the beginning of the 20th century, the avenue was characterized as a tree-lined square and its historical monuments were preserved. Today we see a complete lack of trees, uncharacteristic façades and a great deal of visual pollution as a result of the number of parked vehicles and overhead electrical wiring. It is important to carry out an intervention that seeks a new image for the avenue, requalifying the site through the implementation of tree planting, communal areas and a new road organization, with the aim of improving the environmental comfort of the place and improving the quality of life for those who work and circulate there, with tree planting as a key element in this process of improvement.

Key words:

Landscape, Arborization, Comfort.

INTRODUCTION

[...] As an old city, like so many other old cities in Brazil, Penedo should keep in every street, in every square and in every neighborhood, a standing testimony that speaks with the very eloquence of its presence, of the times, the men and the things that have passed, yes, but that honor the memory of all those who were anonymous and ardent workers of our nationality and our integrity [...] (GOMES, 1942).

"The city of Penedo is considered, along with the towns of Porto Calvo and Alagoas (now the city of Marechal Deodoro), to be one of the "*mater*" cells of the colonization of the lands of Alagoas" (BEZERRA, 2006, p. 4).

The object of this study is the Floriano Peixoto Avenue and its set of monuments, which tell the story of Penedo's urban and architectural evolution with their representative historical and artistic uniqueness. The avenue underwent several changes in its configuration throughout the 20th century, becoming a square and then a plaza, and tree planting has always been an element present in its dynamics and numerous interpretations.

Today, the historic landscape of Avenida Floriano Peixoto is being de-characterized, where lampposts with overhead electric wiring have taken the place of the old trees, and vehicles have taken the place of people, the result of an inevitable process of modernization of the urban space.

Thus, this work proposes a "requalification"[1] of the landscape of that urban space, taking the existing trees as a reference, seeking a new image for the avenue and, above all, improving the environmental comfort of the place, which is currently so arid and full of elements that add no value and only de-characterize the landscape of Penedo's Historic Site.

In addition, the aim here is to propose guidelines for the implementation of places of interaction and a new road organization, as a means of providing a better quality of life for those who work and circulate there and seeking

---

1 Appreciation of the symbolic, historical, socio-economic and cultural character, integrating the various aspects that make up the urban habitat.

integration with the preservation of the historical and artistic heritage of this listed area.

This work also aims to contribute to the enrichment of the discussion about the lack of trees in historical sites, often characterized by the absence of trees on public roads, and consequently by the hot and arid climate.

The methodology used to apprehend the area studied is based on that applied by Kevin Lynch in his book: The Image of the City (1997), which studies the visual quality of the city through the mental image that its inhabitants make of it and will be addressed in this work. The methodological tripod used in this work is based on the components of the environmental image: identity, structure and meaning, cited by Lynch (1997).

The first component of this tripod seeks to read/understand <u>the identity of </u>the area under study in relation to conservation theories that deal with the essence and unity of the cultural asset, defended by theorists such as John Ruskin and Cesari Brandi and which will be addressed in this work [emphasis added].

The second relates the <u>structure of </u>the environmental image referred to by Lynch (1997), which includes the relationship of space between the object and the observer to technical knowledge about urban afforestation, which will also guide the proposal of this work, such as the analysis/reading of the area currently for the implantation of this afforestation and urban equipment [emphasis added].

Finally, the third stage seeks to capture the <u>meaning </u>that the object of study assumes for the environmental image of the observer, which will be sought through popular participation, capturing the urban image of the place through the inhabitants of the city of Penedo. This methodology, which includes popular participation through questionnaires with the population, aims to capture the public image, critical elements and visual qualities of Avenida Floriano Peixoto. As this is a historic landscape, where symbolism and the collective imagination are so present, it also aims to understand how the avenue is perceived by the population and whether there is a demand for tree planting in the area. This is in addition to my own apprehension of that reality, based on my vision as an architect and urban planner [emphasis added].

To this end, the first chapter addresses issues of landscape and urban image through Kevin Lynch's (1997) analysis of the physiognomy of cities, the fact that this physiognomy has some relevance and the possibility of modifying it. We also talked about urban afforestation, the importance and benefits it brings to the urban environment, and technical knowledge about the characteristics of suitable species and how to make them compatible with public roads. Finally, a brief history of gardens in Brazil was developed, focusing on the issue of afforestation in historic sites and the preservation of these gardens in relation to conservation theories. Thus, the first chapter contributes to the development of the methodological tripod proposed through the identity and structure of the environmental image dealt with by Lynch (1997).

The second chapter analyzes the historical development of the city of Penedo, the listed historic site and how the avenue is part of this context, the significance and importance of this road in the history and current dynamics of the city. A more specific analysis of the object of study was also carried out, with historical references to the emergence of the avenue, its architectural ensemble and its evolution and changes over the years. The chapter in question aims to contribute in part to achieving the second element of the methodological tripod by analyzing the structure of the environmental image of the area, which will provide subsidies in relation to the technical issues of the proposal, and the third which relates the meaning of the element studied for the author at this time.

The third chapter applies the methodology of reading/thinking about the area under study and the concepts developed in the previous chapters in order to draw up guidelines for the afforestation of Floriano Peixoto Avenue, so as to achieve the general objective of this work, complementing the third element of

Lynch's (1997) methodological tripod, meaning.

This meaning would be the role of the Avenue in the historic center of Penedo and its great importance within Penedo society, not only artistically, culturally and historically, but also in the economic dynamics of the city.

1 QUESTIONS ABOUT URBAN LANDSCAPE

1.1 THE IMAGE OF THE CITY

Over time, the city grows on itself, acquires an awareness and memory of itself. In its construction, the original motifs remain, but at the same time, the city makes the motifs of its development more precise and modifies them (ROSSI, 2001, p.2).

"As an architectural work, the city is a construction in space, but a large-scale construction, something only perceived over long periods of time" (LYNCH, 1997, p.1). There is always a landscape to be explored, a surprise to be revealed in the sequence of elements that make up the urban landscape. For Kevin Lynch (1997), an isolated building does not represent a significant attraction for the eye, but a group of buildings acquires a power of visual attraction because it is enhanced by the elements that surround it, by the sequences that lead to it and by memories of previous experiences. The image has vast associations and is impregnated with memories and meanings for each individual in the various landscapes that make up a city.

For Gordon Cullen (1983), the elements of a city should have an emotional impact on people, reacting to the contrasts between full and empty, differences in scale, styles, colors and textures, but if the city appears monotonous, uncharacteristic or "amorphous", it will not be fulfilling its mission. "[...] vision has the power to invoke our reminiscences and experiences, with all their corollary emotions, a fact that can be used to create extremely intense situations of enjoyment" (CULLEN, 1983, p.11).

It can be said, then, that a landscape exerts a power of communication with the observer when it arouses some kind of emotion in them. This is what can be observed, for example, when you come across the landscape of a historic site, which reveals in its architectural complex and urban layout the marks of past eras. This is why they are impregnated with the memories and meanings referred to by Lynch, and have the power to arouse an emotional impact in people, according to Cullen (Image 01).

Image 01: Pelourinho: Historical landscape imbued with meaning.

Source: http//www.sherlockholmes.com/imagens/hpm.

Through an analysis of the physiognomy of cities, whether or not this physiognomy has any relevance and the possibility of modifying it, Lynch (1997) suggests a method by which one could begin to deal with visual form on an urban scale, applied to the central areas of three North American cities: Boston, Jersey City and Los Angeles through interviews with some citizens about their image of the environment and a systematic examination of the environmental image provoked in the field by experienced observers.

According to Lynch (1997), all this material would be synthesized through maps and reports that would show the public image of the area, the visual problems and qualities, the critical elements and on the basis of this analysis, continually transformed and updated, a project could be made for the future visual form of the region.

The design of a city can be considered a temporal art, as it is perceived in different ways, at different times and by different people, and its meanings can be inverted or take on new forms according to the perception of the time (LYNCH, 1997).

Thus, the city can be considered as a set of elements imbued with meaning, whose forms are related to those of other sets of products and social practices, adding to the values and meanings that permeate the urban landscape and contributing to an aesthetic quality related to its understanding and appreciation (RAPOSO, 2001). "What transforms any cultural event into art is the quality of the expression that can provoke a deep emotion in the subject. Therefore, what is at stake is the valuation of the expressive quality" (MENDES apud RAPOSO, 2001, p.28).

In this way, the design of a city, the layout of its blocks, streets and urban voids, take on their own forms and are built up as society's values and perceptions change over time and transform the urban landscape (LYNCH, 1997).

It can therefore be said that people and their activities are just as important as the physical part of a city, sharing the same landscape and participating in its formation. According to Lynch (1997), the observer perceives the city in a partial, fragmented way, mixed with their personal knowledge and value, with the various sensations it provokes in them, but the image they grasp is the combination of the perception of all their senses working together at a given moment.

According to Lynch (1997), the city is the product of the action of many agents who are always modifying its form. Only in parts can control be exercised over its growth and form: on the one hand it can remain unchanged for some time, and on the other it is always changing in detail, so the city is the product of a dynamic sequence of phases (Image 02).

Image 02: Penedo's historic center: a product of the dynamic succession of phases.

Source: http//www.casadacultura.org.

This can be clearly seen when you analyze the historical evolution of the avenue shown in the photo above, with its architectural monuments that are records of several different eras and styles and have been modified according to the new concepts and needs of society, which has changed over time.

Through a methodology that examines the visual quality of the city, studying the mental image that its inhabitants make of it, Lynch (1997) focuses mainly on a specific visual quality to classify the urban landscape - the clarity or "legibility" of the urban setting, showing how this concept could be used to give the city a new shape. "A legible city would be one whose neighbourhoods, landmarks or streets were easily recognizable and grouped into a general pattern" (LYNCH, 1997, p.3).

Although clarity or legibility is by no means the only important attribute of a beautiful city, it is something that takes on special importance when we consider environments on the urban scale of time and complexity (LYNCH, 1997, p.3).

He also explains that the legible image of an environment plays an important social role, as it can serve as a reference, an organizer of the culture of a given society. An individual can create a harmonious relationship between him and the environment around him when an image seems recognizable or orderly, by provoking in him a feeling of emotional security and orientation, through an immediate sensation transmitted by the image or by the memory of past experiences. "An admirable landscape is the skeleton on which many primitive races erect their socially important myths" (LYNCH, 1997, p.5).

According to Lynch's concepts, the city of Brasilia could be considered a legible city, with its completely orthogonal and symmetrical layout, drawn from a central axis and divided into orderly sectors. In this way, it would be easily organized in general by the inhabitants or people who already knew its design, just as it could be considered an unreadable city for an outsider who couldn't find their way around the monumental scale of this city, which was unknown to them (Image 03).

In the same way, it could be said that a resident of a medieval city, with its winding alleys and non-orthogonal layout, would easily get around because they already know it, while an outsider would find themselves lost in front of this irregular and unfamiliar layout, thus proving that the concept of legibility can be relative to the circumstance (Image 04).

Image 03: Urban layout of Brasilia: the orthogonal is not always the most legible.

Source: http//www.sbhmat.com.br.

Image 04: Medieval town: irregular layout may be legible to local inhabitants. Source: http//www.stgema.com/gallery/photos/lucca.

However, Lynch (1997) goes against the importance of the concept of legibility when he argues that the human brain is easily adaptable, and that even in a totally disordered environment it is possible for an individual to be able to orient themselves and organize themselves, which is not uncommon in completely disorganized urban centers in Brazilian cities.

The same could be observed in the case of Brasilia, where at first an outsider would be disoriented by the similarity of its blocks and the monumentality of its streets, but over time would be able to orientate themselves and adapt to the logic of the city's design. In the same way, a visitor to a medieval city, after some time, would be able to adapt and find some logic in its irregular layout.

Urban planners are basically interested in the external agent who creates the urban image, but for Lynch (1997), the most important thing for these urban planners when idealizing an environment that will be used by various individuals are the consensual images of a homogeneous and significant group of observers who interact in the same reality. This is an important observation by Lynch (1997), since there is a current trend towards popular participation in urban decisions and interventions, such as the Participatory Master Plan, which seeks to build inclusive and democratic cities.

This methodology, which includes popular participation using the urban image created by the inhabitants of a city, through interviews with selected citizens, could be applied to the object of study of this work. This is a significant avenue in the historic center of the city of Penedo and the aim would be to capture the public image of the area, the critical elements and the visual qualities, since symbolism and the collective imagination are so present in a historic landscape. Then, based on this analysis, to propose guidelines for a new visual form for the avenue, with the implementation of tree planting on its public sidewalks, as a new element of landscape composition and transformation of the urban image.

"An environmental image can be broken down into three components: identity, structure and meaning" (LYNCH, 1997, p.9). The identity of an image relates to the meaning of individuality or differentiation from other things, its structure includes the relationship of space between the object and the observer and other objects, while that object assumes some Meaning for the observer, whether practical or emotional (LYNCH, 1997). However, he explains that, for the analysis of the urban environment, consensual images of meaning tend to have less consistency than perceptions of identity and structure, because individual meanings are so diverse, even when their form can be easily communicated, that it seems impossible to separate meaning and form.

According to the same author, there is also a visual quality that organizes the mental image - imaginability, which is the ease of constructing clearly recognizable, orderly and useful mental images of the environment, through the characteristics of a physical object.

A highly "imaginable" city would invite the senses to greater observation and participation, perceived over time as a model with many distinct parts clearly interconnected (LYNCH, 1997) (Image 05).

The statue of Christ the Redeemer in the city of Rio de Janeiro has, according to Lynch's ideas, the characteristic of creating easily identified and structured mental images, as it is completely integrated into the city's landscape and invites the senses to pay greater attention to its beauty and monumentality.

"Among its many roles, the urban landscape is also something to be seen and remembered, a set of elements that we hope will give us pleasure" (LYNCH, 1997, [n.d.]).

Image 05: Christ the Redeemer - clearly identified image. Source: http//everttaube.info/images/rio_janeiro.

The importance of green mass in the urban landscape can be added here as an element that contributes to the visual quality of cities, and can be considered a significant structuring component of an environmental image instilled with identity and meaning, and can serve as an organizer of the mental image that its inhabitants make of it, establishing a relationship between the individual and the formation of the urban image.

1.2 URBAN AFFORESTATION

This land, Sir, seems to me that from the point that we see most to the south, to the other point that comes to the north, which we have seen from this port, will be so large that there will be twenty or twenty-five leagues of coastline. Along the sea it has some large barriers, some red and others white; and the land above is all flat and full of large trees. From end to end, the whole beach [...] is very flat and very beautiful. From the sea, it seemed to us to be very large, because with our eyes wide open, we could see nothing but land and trees - land that seemed very extensive to us (CHANDEIGNE, 1992, p. 164).

As Mascaró (2004, p.65) shows, "The tree is the most characteristic plant form in the urban landscape, which has been incorporated into a close relationship with architecture throughout history". This important element of landscape composition also contributes to achieving a pleasant urban "ambience"[2] , with its multiple functions for environmental balance.

Trees play a vitally important role in the quality of life in urban centers. Urban trees have a direct effect on the climate, air quality, noise levels and the landscape, as well as being a refuge for the remaining fauna in cities (CURSO...,1995) (Image 06).

---

2 It refers to the quality of the urban environment influenced by its morphology: the width of streets and the height of buildings that define their profiles, as well as tree planting which, as a component of this landscape, contributes to improving environmental balance.

Image 06: Afforestation contributes to a pleasant urban environment.

Source: http//www.skscrapercity.com/showthread.php?t=230981.

Characteristics of the urban environment, such as soil sealing due to paving and construction, reduction of vegetation cover and atmospheric, water, visual and noise pollution, make the standard of the urban environment much lower than that necessary for adequate human living conditions. "Vegetation, however, through its ecological, economic and social functions, can play an important role in improving the lives of urban populations [...]" (CURSO..., 1995, p.7).

According to Mascaró (2004), trees protect buildings from unwanted sunlight, reducing energy consumption during the hot season by filtering the sun's rays. They form barriers and channel ventilation, arranged singly or in groups, where the type of foliage and density of the canopy must be considered for the desired effects. Acting on climatic elements in urban microclimates, it also helps to control air temperature and humidity, the action of winds and rain and to mitigate air pollution. "These forms of use vary according to the type of vegetation, its size, age, period of the year, forms of plant association and also in relation to buildings and their urban enclosures" (MASCARÓ, 2004, p.66).

Trees and other plants intercept, reflect, absorb and transmit solar radiation, improving the air temperature in the urban environment, according to the characteristics of their foliage and branching (GREY AND DENEKE apud CURSO..., 1995). "The short-wave radiation that hits the leaves will be partially transmitted as diffuse radiation, because the leaf is not opaque to solar radiation" (MASCARÓ, 2004, p.66). According to him, vegetation also modifies wind characteristics, which also regulate human comfort. In summer, the action of the wind increases evaporation and in winter it cools the air. "The movement of air regulates thermal sensation, as it stimulates evaporation and heat loss by convection" (MASCARÓ, 2004, p.67) (Image 07).

The presence of trees in cities also helps to reduce air pollution, with their considerable potential for removing polluting particles and gases from the atmosphere. "Tree leaves can absorb pollutant gases and trap particles on their surface [...] However, the capacity to retain and tolerate pollutants varies between species [...]" (SCHBER apud CURSO..., 1995, p.11).

Image 07: Trees as regulators of the urban microclimate.

Source: http//www.skscrapercity.com/showthread.php?t=230981.

The many benefits that afforestation can bring to the urban environment are therefore clear, as is the need for a study of the characteristics of the species to be used and adequate technical-scientific planning for its implantation in the urban space with its multiple purposes of housing, infrastructure, circulation, services and production.

In addition to these considerations, there is the important psychological role of trees for human well-being and the aesthetic quality they play in the urban landscape as an important element of landscape composition. Thus, characteristics such as size, shape, texture, structure and color should be analyzed so that they establish a harmonious relationship with the landscape between the parts of the whole and between the parts of the whole and the whole (Image 08).

When selecting the species to be used in urban afforestation, it is necessary to know the primary needs and requirements of the vegetation, such as those related to soil, water, light and the local environment, as well as their characteristics and behavior (CURSO..., 1996). Favorable aspects of the species should be taken into account, as their efficiency has been proven in practice:

Image 08: Trees as an element of landscape composition.

Source: http//www.skscrapercity.com/showthread.php?t=286971.

a)   Origin - Give priority to species native to the region and already adapted to the area;

b)   Size and shape - The species chosen should have a single stem and a well-defined canopy, with a size compatible with the location;

c)   Rootstock - This should be pivoting and deep, avoiding damage to urban infrastructure;

d)   Growth - Choose species with medium to rapid growth, avoiding depredation or death of the plant;

e)   Fruits and Flowers - Choose those that produce small fruits for birds, avoiding those with heavy fruits and possible accidents from falling, and flowers that remain on the plant for a long time and have bright colors;

f)   Leaves - Give preference to species with semi-deciduous foliage, which remain on the plant for longer, or to those that are persistent, and observe the scale relationship between the foliage of the species and that of the place where it will be planted;

g)   Resistance - Species that are resistant to pests and diseases, as well as to natural weather conditions, should be chosen, avoiding those with toxic principles.

Adding to this are the sensory qualities of the species, in the case of those that convey sensations, such as olfactory with species that exude aromas, auditory with those that provoke some auditory stimulation, through the foliage and visual with the choice of species that have visual qualities, whether through their shape, leaves or flowers.

In fact, all these aspects must be observed, since, due to the conditions of the urban environment, trees behave differently than if they were in their natural environment, thus ensuring their adaptation, survival and development in the place to be implanted.

Along with this analysis of vegetation, there is a need for adequate knowledge of the characteristics and conditions of the urban environment, taking into account factors such as local aspects, available space, type of occupation and traffic and environmental characteristics, and there must also be compatibility between the tree planting and the electrical system, water supply, sewage, signs and buildings (CURSO...,1996). According to the same author, some measures should be observed when implementing urban tree planting, such as minimum setbacks from the curb, clearances between the tree canopy and the electricity network, and maximum tree heights according to size (Image 09).

However, what has been observed in several Brazilian cities is the lack of planning for the implementation of tree planting, from planting to proper maintenance and pruning, causing interference in urban equipment and infrastructure, inconvenience to the population, and the implementation of measures that harm the development or lead to the death of tree species, such as drastic pruning and root cutting (Image 10).

Image 09: Interference from trees on the overhead electricity network.

Source: http//www.skscrapercity.com/showthread.php?t=286971.

Image 10: Damage to urban facilities caused by poorly planned afforestation. Source: www.infobibos.com/.../ArborizaçâoUrbana.htm.

"It is known that the urban climate differs considerably from the natural environment. The temperature range, the rainfall regime, the water balance, the humidity of the air, the occurrence of windstorms and the artificial luminosity of cities [...]" (SANTAMOUR apud CURSO..., 1995, p.20) all need to be taken into account. According to the same author, urban soil is almost always compacted and polluted by solid waste and residential and industrial waste, preventing full development or causing damage to trees.

Therefore, we can see the importance of planning tree planting for the urban environment, with characteristics so adverse to those of the natural environment, in order to obtain the benefits that urban trees provide, such as environmental balance, human comfort and landscape quality, and to establish a harmonious relationship and a perfect interaction between vegetation and built space.

1.2.1 Trees on public roads

The street is the urban space for public use whose function is to organize and relate the architectural facts in the urban fabric. [...] Its environment is one of movement, but also a space to be experienced [...] The width of the street, the height and characteristics of the bilateral pianos, as well as the presence of vegetation affect the urban ambience [...] (MASCARÓ, 2004, p.87).

The afforestation of a city's streets, when properly planned and managed based on technical and scientific foundations, brings countless benefits to the population. Among the possible socio-economic benefits of street tree planting are the influence on the health and physical and mental well-being of the population and the enhancement of property values. In addition, trees on public roads perform an ecological function, in the sense of improving the urban environment, and a landscape function, in the sense of beautifying public roads and consequently the city (MANUAL..., 1996).

According to the same author, another important function of street trees is that they act as an ecological corridor, linking the city's free vegetated areas, such as squares and parks. In addition, on many occasions, the tree in front of the house gives it a particular identity and allows residents to have direct contact with a significant natural element, considering all its benefits and even greater interaction between people, since the shading is inviting (Image 11).

As we have already seen, the presence of vegetation in urban areas is fundamental, bringing quality of life to its inhabitants. Trees planted on public sidewalks provide aesthetic quality and help to keep the temperature down, especially in summer, when the scorching sun mistreats passers-by. They also reduce noise pollution and temperature, release oxygen into the atmosphere, increase air humidity and absorb carbon dioxide (COURSE..., 1996).

However, there are many problems caused by the confrontation of unsuitable trees planted on public sidewalks with urban equipment, such as overhead electric wires, pipes, gutters, sidewalks, walls and lampposts, most of which cause improper and harmful management of trees. In this sense, it is essential to consider the need for constant and appropriate management specifically aimed at street tree planting. This management involves the stages of planting, handling the seedlings, pruning and the necessary removals (COURSE...,1996) (Image 12).

It is necessary to consider the need for specific municipal legislation, administrative measures aimed at structuring the competent sector to carry out the work, considering, fundamentally, qualified manpower and appropriate equipment, as well as the involvement of the population in general, in the process of afforestation or reafforestation of the city (CURSO..., 1995, p.22).

Image 11: Ecological corridor connects green areas.

Source: http://www.edsonlimaimoveis.cim.br/images/Praça%20Tamandaré.

Image 12: Lack of afforestation due to planning difficulties.

Source: http//www.skycrapercity.com/showthreead.php?t=230981.

The preservation of urban trees is already the subject of specific legislation in some cities, such as São Paulo and Curitiba, with the aim of keeping the trees on public roads in a desirable condition. One of the most important conservation practices is pruning, which seeks to ensure that the functions performed by trees are in harmony with the urban environment, guaranteeing their vitality and people's safety, as well as carrying out pre-planned planting and using appropriate techniques to guarantee the trees' development.

Proper urban planning should provide for a central green bed in sufficiently wide streets or a grassy strip or bed along the sidewalks, between the curb and the buildings, reserved for planting trees. This makes it possible to use species with shallow roots and any damage is restricted to the grassy strip, allowing for greater absorption and penetration of rainwater, and consequently greater soil respiration (WYMAN apud CURSO...,1995), however, in areas that are already consolidated, this practice becomes almost unfeasible (Image 13).

Image 13: Width of the road allows for central tree planting. Source: http//www.skycrapercity.com/showthreead.php?t=303906.

As has already been pointed out, there are aspects to be observed when planning urban tree planting. If trees are to be planted on public roads, even greater care must be taken, by analyzing the vegetation that adapts well to the climate and soil conditions, that has favorable characteristics for the urban environment and is preferably native to the region, and by analyzing the site to be afforested, taking into account aspects such as urban infrastructure, type of occupation and environmental characteristics. It is also essential to match the

22

size of the tree, the width of the street and the electrical wiring network (MANUAL..., 1996). According to the author, compatibility should be done as follows:

a)   Narrow streets (less than 7m) and wide sidewalks (more than 2m) - plant small and medium-sized species on the side where there are no wires. Under the wire, plant small species, alternating with those on the other side of the street (Image 14).

b)   Narrow sidewalks (less than 2m) and wide streets (more than 7m) in residential areas - plant only on the side where there are no wires, 50 cm outside the sidewalk, if there is no setback between the building and the sidewalk, small and medium-sized species, if there is a setback, small species can be planted on the sidewalk (Image 15).

c)   Wide sidewalks (over 2m) and wide streets (over 7m) - Plant medium-sized species on the side without wires and small-sized species on the side with wires (Image 16).

d)   Wide sidewalks (over 2m) and wide streets (over 7m), without wiring or with underground wiring - Plant medium-sized species on both sides and medium and large-sized species when there is a central flowerbed (Image 17).

Image 14: Medium-sized species without an electricity grid.

Source: http//www.skycrapercity.com/showthreead.php?t=303906.

Image 15: Small trees on the sidewalk in a residential area.

Source: http://www.casaecia.arq.br/IMAGES/nerium_oleander.

Image 16: Small trees with electricity grid.

Source: http//www.copel.com/.../carcterísticas.htm.

Image 17: Medium and large trees on a wide street.

Source: http//www.skycrapercity.com/showthreead.php?t=303906.

Also for compatibility with the electrical system, it recommends observing the minimum distances between trees and the electricity grid and overhead telephone networks, and checking the location of underground networks.

As for the type of occupation, for commercial streets, he recommends planting away from the curb, to avoid damage caused by vehicles, and the use of metal grids over the pit, to avoid trampling by pedestrians and the consequent compaction of the soil (Image 18).

Image 18: Metal grating to prevent scattering.

Source: http//www.ufrrj.bb/.../it/de/acidentes/baciaurb.htm.

Compatibility is therefore possible, provided that the implantation of trees on public roads and other urban facilities is planned in an integrated manner, with due knowledge of the species to be used and the characteristics of the site to be implanted.

It's also worth pointing out that tree planting on public roads needs to be appropriate, taking into account the variations that may exist in relation to the size of streets and sidewalks and the configuration of the urban layout. In the case of some historic Brazilian cities, for example, with a layout that often has narrow, winding roads with no distance from the buildings, the compatibility of trees with these conditions must be even better studied and adapted to ensure the successful selection of the species with the characteristics favorable to its implantation and integration with the landscape, as we will see below.

1.3 ARBORIZATION IN HISTORIC SITES

1.3.1 A brief history of gardens in Brazil

[...] And in the middle of that barren and fruitless sand, he planted a garden, and all the varieties of fruit trees that are found in Brazil, as well as many that came from different parts, and the strength of a lot of other fruitful soil, brought from outside and creeping boats, and a lot of manure, made the place as well conditioned as the best fruitful land [...] (PISO apud ALCIDES, 2005, p.135).

The formation of Brazilian colonial society can be seen as a "[...] transposition of laws, customs and spiritual equipment from the metropolis" [...] (DELPHIM, 2005, p.12), which placed little value on or disregarded local cultures and customs. This difference in cultures was always present in this formation, which highlights the contrasts between two realities, the urbanized and refined European and the primitive and rustic Brazilian. In the same way, the garden was transposed along the lines of European gardens, being transformed in parallel with the new Brazilian society that was being formed (DELPHIM, 2005).

"The context of the exploration of the American lands in the 16th and 17th centuries was characterized by an experience of global research in which the discovery and knowledge of the other intensified [...]" (ALCIDES, 2005, p.20). The descriptions of Brazilian flora, from the arrival of Pedro Alvares Cabral, reflect the impact that the exuberance and diversity of botanical specimens still unknown in Europe had on the colonizers, which can be seen in the chronicles of travelers who reported their experience of visiting the unknown land (ALCIDES, 2005) (Image 19).

At the time of the discovery, the Portuguese witnessed the beginnings of an agricultural activity in Brazil, which they adapted to their own interests, and which contributed most to the transformation of the natural landscape, with rare cases of gardens in the country until the Empire. Thus, they exploited native species and gradually introduced exotic species, such as sugar cane, which was the first and most important crop brought to the country and also the one that most devastated the natural vegetation cover (DELPHIM, 2005).

Image 19: The Indian and Brazilian nature under the European gaze.

Source: http//www.ifch.unicamp.br/ihb/MaxIndios.

According to the same author, the house-large-room system of the new rural sugar society and its architectural ensemble, "[...] was associated with a landscape treatment defined by the orchard, decorative plants and attempts to ennoble the property [...]" (DELPHIM, 2005, p.13) (Image 20).

Image 20: The sugar mill and its architectural and landscape complex.

Source: http//www.cepa.it.usp.br/energia/energi1999/Grupo1B/Image14.

With the decline of rural patriarchy and urban growth, the big house decreased in size and complexity, took on the urban form of townhouses and slave quarters, and was surrounded by mocambos and tenements. In this context "[...] the square replaces the mill, but only little by little [...] the gardens, the so-called public sidewalks and the squares remained for a long time enclosed with iron bars, limited to the use and enjoyment of people of one class [...]" (FREIRE apud DELPHIM, 2005, p. 13).

According to Delphim (2005), Brazilian gardens, from the 16th, 17th and 18th centuries, were an artificial landscape that sought to ennoble private, rural and urban property related to the orchard, with imported fruit trees such as avocado, lemon, mango and sapodilla trees, and to the areas where domestic animals were raised, from where he believes the terms "quintal", "sitio" and "roça" came from, related to the Portuguese association of the garden with the vegetable patch and the orchard.

"With regard to the organized landscape of cities, during the colonial period we highlight the urbanization of Recife and Olinda, in the first half of the 17th century, and the landscaped squares of Rio de Janeiro from 1753 onwards [...]" (DELPHIM, 2005, p.14), especially the Passeio Pùblico in Rio de Janeiro, which for some time was the only landscaping project carried out for the enjoyment of the city's population (DELPHIM, 2005) (Image 21).

The presence of the Dutch in Brazil under the patronage and encouragement of Count Joâo Mauricio Nassau-Siegen, who designed the Parque de Friburgo, a large green space in the middle of the 17th century in the city of Recife, and his important historical, political, scientific and artistic research, is a milestone in the botanical history of Brazilian nature, with important studies and records of it (NOGUEIRA apud ALCIDES, 2005) (Image 22).

Image 21: The geometric layout and tree-lined rectilinear boulevards of the Passeio Publica. Source: http//www.upload.wkimedia.org/wikipedia/commons/thumb/7/73/InauguraçâoPortâo- finsXVII.

Image 22: Frans Post's sketch of Fribourg Park. Source: http//www.dw-world.de/image/o,1116882_1,00.

With the transformation of Brazil's cultural characteristics following the transfer of the Portuguese Court to the country in 1808, the aspiration was for an empire along European lines, and a second moment of transformation of the Brazilian urban landscape began. The opening of the ports and the arrival of the French Artistic Mission were part of a set of measures to encourage cultural activities in the country through French artists and artisans who attributed an "[...] ideology based on the double rupture with the colonial Portuguese past and with the aspects inherent to the Brazilian land and the society it had formed [...]" (CADERNOS... apud DELPHIM, 2005, p.15).

According to Delphim (2005), with the loss of the monopoly of the Orient, Portugal adopted a policy of producing species in Brazilian lands and creating botanical gardens such as those in Belém in 1796, Salvador in 1803, Rio de Janeiro in 1808, Olinda in 1811 and Ouro Preto and São Paulo in 1825, in 1808, Olinda in 1811, Ouro Preto and São Paulo in 1825, which aimed to acclimatize useful plants and implement the trade in species in Europe and the introduction of new species to rival those from the East (Image 23).

Image 23: Rio Botanical Garden: Neoclassicism transferred to the language of the gardens.
Source: http//www.baixaki.ig.com.br/Imagens/wpapers/BXK18321_dsc04980-riojaneiro-jardim-bot800.

"In Rio de Janeiro, English and French influence prevailed, the latter officially introduced by the French Artistic Mission. The projects of Grandjean de Montigny and Auguste François Marie Glaziou stand out from this period" (DELPHIM, 2005, p. 19). Through this influence, we can see the presence of neoclassicism, imposed at the time as official art, in the design of these gardens, where we can see the search for rigorous composition with the use of symmetry and formal austerity in the characteristics and forms they acquired. Later, the Sâo Paulo Botanical Garden still shows in its geometric layout the formal vocabulary imported and transferred to the composition of the gardens (Image 24).

It was in this context, through expeditions led by foreign naturalists, that studies of Brazilian nature began, contributing to the selection and introduction of various native wild plants for cultivation and landscaping in private gardens (DELPHIM, 2005).

Image 24: Influence of the French garden on the design of the Jardim Botànico de Sào Paulo.

Source: http//www.scielo.br/img/fbpe/rbb/v24n4s0/9480f1.

On the other hand, from the 18th century onwards, during the government of Dom Joâo VI, with a policy that favored the urban bourgeoisie, with the cultivation of coffee and the consequent immigration to São Paulo and Rio de Janeiro, according to Delphim (2005), this was a new factor in the transformation of the built landscape, with a considerable increase in private and public buildings. This period, marked by the development, wealth and rise of the bourgeoisie and the consequent importation of elitist European vocabulary, reflects the European influence on the urbanism and architecture of the time, as well as on the gardens that were produced in Brazil.

In the 19th century, with the changes to residential plots, now with side setbacks where gardens were built, the walls gave way to iron railings and the gardens separated from the vegetable garden and orchard were left to French gardeners who used European species and promoted a taste for the picturesque. From then on, there was an intense trade in foreign plants, which could only be bought by an elite who showed off their gardens on the sides or in front of their houses, with mainly English and French influences (DELPHIM, 2005) (Image 25).

Image 25: Gardens in side setbacks of residential plots.

Source: http//www.dci.ufscar.br/parqscar/Image138.

Meanwhile, the native or non-European plants seen in mocambos and poor houses, with indigenous or African and Asian species useful for domestic use, were despised or considered "low people's plants" or "macumba plants" (FREIRE apud DELPHIM, 2005, p. 21).

According to Delphim (2005), until 1900, the re-Europeanization of the Brazilian urban landscape and the adoption of Eclecticism and Art Nouveau, reproduced mainly in the country's major cities in private and public buildings, characterize an era that favored an artificial landscape that ignored the primitive Brazilian culture.

However, from the 1920s onwards, there was a moment of rupture, which would later be called Modernism, in which intellectuals, particularly artists and writers, denounced and criticized an elite and bourgeoisie that aspired to be "civilized" and a country that was considered an extension of European culture.

This rupture led to profound changes in Brazilian cultural and social life and an end to the sense of inferiority in relation to Portugal (DELPHIM, 2005).

From then on, the search for a nationalist expression determined the very originality of Modernism, where local references were sought and Brazilian tropical nature was valued. This is where the figure of landscape designer Roberto Burle Marx[3] comes in, who saw the garden as an adaptation to the natural and ecological environment, studying and using species native to Brazil and seeking to meet the natural needs of society (MARX apud DELPHIM, 2005) (Image 26). "From the 1930s onwards, the history of the Brazilian garden is linked to the achievements of modern architecture and the work of Burle Marx" (DELPHIM, 2005).

Image 26: Burle Marx: Native species of the Brazilian caatinga in Euclides da Cunha Square, Recife, PE.

Source: http//www.ceci.-br.org/novo/revista/images/articleimages/CT-2005-18-image001.

According to Delphim, at the beginning of the 1980s, interest in the treatment of historic gardens with criteria similar to those used in the preservation of other cultural assets emerged in Brazil, with the Fundaçâo

National Pro-Memory in the Botanical Garden of Rio de Janeiro. The aim was

---

3 Roberto Burle Marx (1909-1994) was an innovative Brazilian landscaper who rethought the whole concept of aesthetics and landscape art. He was concerned with ecological balance and constant observation, especially of Brazilian plant species. He was an observer and consequently a discoverer of new species, which therefore bear his name, such as: Calathea burle-marxii, a precedent from Càceres, Mato Grosso do Sul.

to define the issue of monuments, sites and natural landscapes, which had not in fact been dealt with by the National Historical and Artistic Heritage Institute (IPHAN), although it had been established by Decree-Law 25/37[4] . The issue of the surroundings, of natural assets in cultural environments, was incorporated into the debates on the isolated built asset, enriching the vision of landscape heritage and giving rise to a wide range of activities by including professionals linked to botany, landscaping and gardening.

1.3.2 Preserving historic gardens

The plant is form, color, texture, aroma; a living being with needs and preferences, with its own personality. [...] You can think of a plant as a brushstroke, or as an embroidery stitch - without ever forgetting that it is an individual. You should always make it look as much like itself as possible (FLEMING apud SIQUEIRA, 2001, p.31).

In 1981, the International Committee for Historic Gardens and the International Committee on Monuments and Sites/ International Federation of Landscape Architects - ICOMOS/IFLA drew up the Florence Charter[5] , on the protection of historic gardens, following the principles of the Venice Charter for the Conservation and Restoration of Historic Monuments approved in 1964 in Venice-Italy and establishing specific standards for historic gardens

---

5 DECREE-LAW № 25, of November 30, 1937 - Organizes the protection of the National Historical and Artistic Heritage. CHAPTER I - The National Historical and Artistic Heritage. Art. 1 - The National Historical and Artistic Heritage is made up of all the movable and immovable assets existing in the country and whose conservation is in the public interest, either because of their connection to memorable events in Brazilian history, or because of their exceptional archaeological or ethnographic, bibliographic or artistic value.
§ Paragraph 2 - Natural monuments, as well as sites and landscapes which it is important to conserve and protect due to the remarkable features with which they have been endowed by nature or enhanced by human industry, are equivalent to the assets referred to in this article and are also subject to listing. BRAZIL. Decree-Law No. 25 of November 30, 1937, **which organizes the protection of the National Historical and Artistic Heritage**. Available at: <http://www.lei.adv.br/25-37.htm> Accessed on: September 25, 2007.
6 Meeting in Florence on May 21, 1981, the International Committee on Historic Gardens and ICOMOS/IFLA decided to draw up a charter on the protection of historic gardens, named after the city. This charter was drafted by the committee and registered by ICOMOS on December 15, 1982, with the aim of complementing the Venice Charter in this particular field.
Article 1 - A historic garden is an architectural and plant composition which, from the point of view of history or art, is of public interest. As such, it is considered a monument.
BRAZIL. National Historical and Artistic Heritage Institute (IPHAN). **Letter from Florence**. Available at: <http://portal.iphan.gov.br/portal/baixaFcdAnexo.do?id=252> Accessed on: September 25, 2007.

(DELPHIM, 2005).

For Delphim (2005), some basic concepts of conservation theory should be considered when intervening in a historic garden, such as: values, integrity, authenticity, legal protection and surroundings.

The intrinsic values of a cultural property refer to the physical aspect of the property, including material, conservation, design, location and surroundings. "The essential material represents the intrinsic value of the property and is the support for historical testimonies and associated cultural values, past and present" (DELPHIM, 2005, p.28). The author goes on to mention the extrinsic values of the property, which range from historical, ethnic, symbolic and artistic to social and economic. "A historic garden is a cultural asset that presents these values as well as other physical, natural, cultural and environmental values which, throughout different phases of evolution, have undergone transformations and acquired new and dynamic meanings" (DELPHIM, 2005, p. 28).

According to the same author, the aim of preservation is to safeguard the quality and values of the asset (intrinsic and extrinsic), protecting the essential material and safeguarding its integrity and authenticity for future generations.

According to Delphim (2005, p.29), "Integrity refers to how complete the property is and how much it preserves the balance between the various component elements" of a historic site. He is talking about the inevitable transformations that can take place in cultural property, which are part of the "historical stratigraphy", and which can de-characterize it, or modify its original appearance with amputations or additions, breaking the unity of the property. The interventions proposed for a historic garden will then be based on this new appearance and carried out within the new boundaries (DELPHIM, 2005).

The authenticity of this property refers to the degree of originality of the different elements that make it up, and should reflect the important phases of its evolution, from the moment of its creation to its current form, but never privileging one phase or another, and Delphim (2005) warns that this authenticity would be threatened by the replacement of original elements and "historical strata".

As far as legal protection is concerned, it says that in order to ensure preservation, "historic gardens must be the object of legal protection, in the form of registers, inventories and listings" (DELPHIM, 2005, p. 30), which can be carried out by the National Historical and Artistic Heritage Institute - Iphan, or by other legally constituted institutions according to the model of federal, state or municipal legislation.

Finally, the concept of conserving the surroundings, which is the area that complements the protection of a listed property, aims to preserve the ambience of the listed property. "In the specific case of gardens, consideration should be given to environmental changes that influence lighting, ventilation, [...] microclimate, etc., with restrictions on new [...] modifications that could harm the ambience of the property" (DELPHIM, 2005, p. 31).

Referring back to restoration theorists in 19th century Europe, we can see the emergence of theories that generated the more contemporary ones, with the figure of the French architect Eugéne Emmannuel Viollet-le-Duc with his well-known and contradictory definition that "restoration is the establishment of a complete state that could never have existed" (ANDRADE apud D'ASSUMPÇÂO, 1995, p. 53). 53), when he defends the "stylistic recomposition" of the monument in order to achieve its ideal form, going against the valorization of its historicity, as he disregards the incorporations that the monument has undergone over time and goes beyond the artist who thought it up, already judging that the work could be better than it was thought (D'ASSUMPÇÂO, 1995).

On the other hand, there is some similarity with the contemporary theories cited by Delphim (2005), in another current led by the Englishman John Ruskin, who already established that the aesthetic value should be incorporated into the cultural asset. For him, the marks of time present in monuments should be respected, so any intervention would take away all their authenticity, allowing only conservation operations, and in the case of ruins he advocated that they remain untouched, which shows a position totally contrary to that of Viollet-le-Duc (ANDRADE apud D'ASSUMPÇÂO, 1995).

Back in the 1950s, the Italian Cesari Brandi, in his book "Teoria del Restauro" (Restoration Theory), introduced the dimension of time in restoration, which defines the extent to which restorative intervention can be carried out, considering the work of art from the time it was made to the time it will undergo intervention, which brings it closer to the basic concepts of conservation theories cited by Delphim (2005).

Brandi also argues that the intervention should only be applied to the material, to the physical part, not damaging the image of the work, and that it can also be subject to future interventions or even eliminated and "should be easily recognized without breaking the potential unity of the work" (OSTERMANN apud D'ASSUMPÇÂO, 1995, p. 61). 61), which brings us to the so-called intrinsic and extrinsic values[6] and preservation of the cultural values associated with the property, cited by Delphim (2005), and also to the identity or meaning of the environmental image referred to by Lynch (1997), as seen in the previous section.

Lynch's methodology (1997), which studies the visual quality of a city through the mental image its inhabitants have of it, will be used here due to the importance of capturing this public image of the area under study in this work, an avenue located in a historic site, so that the guidelines proposed for the new visual form of the avenue with the use of a landscape element, the tree, do not de-characterize and preserve the urban image that is part of the collective

---

6 We observed in the landscape of Avenida Floriano Peixoto, the object of this study, its intrinsic value associated with the historical testimonies and cultural values, past and present, of its urban ensemble and also its extrinsic value, related to the artistic, social and economic values of Penedo society.

imagination and history of the city of Penedo.

In addition to the components of the environmental image: identity, structure and meaning, cited by Lynch (1997), for the methodology that will be used here, the conservation theory that deals with the essence and unity of the cultural asset, defended by Brandi, will help in the apprehension of the identity and meaning of the area under study. The technical knowledge on urban afforestation brought up in section 1.2 relates to the structure of the environmental image referred to by Lynch (1997), which includes the relationship of space between the object and the observer and will contribute to the technical aspect of the proposal. And finally, the meaning that this object assumes for the environmental image of the observer, which will be achieved through popular participation, capturing the urban image of the inhabitants of the city of Penedo and the reading/appreciation of the area carried out by the author.

2 THE CITY OF PENEDO

2.1 A BIT OF HISTORY

From the Northeast to Minas Gerais runs a kind of axis, an imaginary line, which not by chance follows the course of the river of national unity, the river Sâo Francisco. Brazil has to return to this axis whenever it doesn't want to forget that it is Brazil (LIMA apud SA; BRASIL, 2005, p.13).

The municipality of Penedo is located in Microregion 121, in the south of the state of Alagoas, in the physiographic zone of Baixo Sâo Francisco. It lies at 10°17'24" south latitude and 36°35'60" west longitude. Its boundaries are: to the north, the municipality of Coruripe; to the south, the river Sâo Francisco; to the east, the municipalities of Piaçabuçu and Feliz Deserto; to the west, the municipality of Igreja Nova (BEZERRA, 2006) (Image 27).

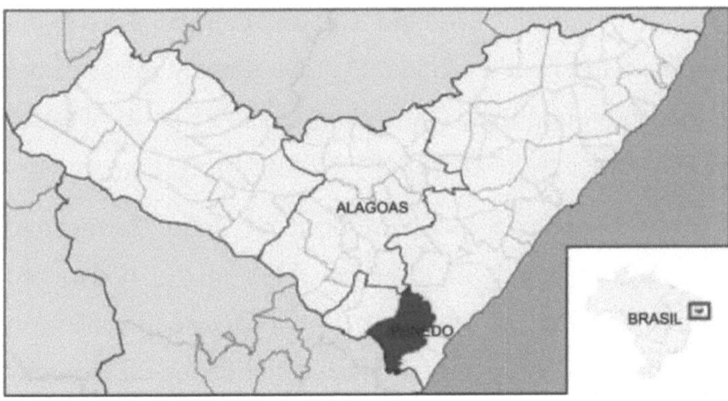

Image 27: Location map of the municipality of Penedo.

Source: Archive of the Department of Planning, Industry and Commerce (SEPLANIC) - Penedo.

With a total area of 688km^2 , Penedo is part of the Atlantic plateau. It has no mountains, but predominantly tablelands, varzeas and hillsides. The municipality is bordered by the Sâo Francisco river and has other tributary rivers, such as the Marituba river, the Piaui river, the Boacica river and the Perucaba river. Several islands form along the length of the Sâo Francisco, including the island of Sâo Pedro and the island of Gado, which are important for the local economy. The vegetation cover is sub-caducifolia forest with valleys running through it. The climate is hot and humid in summer and cold and humid in winter, with an average annual temperature of 26°C and average annual rainfall of 1,600mm. The region's natural mineral wealth includes stone for construction and for making cobblestones. The municipality's main geographical features are its rivers and the rocky terrain (BRASIL, 2002).

According to Bezerra (2006, p. 4), "the town of Penedo is considered, along with the towns of Porto Calvo and Alagoas (now Marechal Deodoro), one of the 'mater' cells of the colonization of the lands of Alagoas". Around 1501, according to historians, on an expedition to the continent, Américo Vespùcio discovered the bar of the great Sâo Francisco and named the river after the Saint of Assisi due to a religious custom, which already appears with this denomination on the oldest map of Brazil, dated 1502, by D. Alberto Cantino.

On the banks of the river Sâo Francisco lived the "Caetés" Indians, allied to the "Abacoatiaras" who lived on the river's islands. Despite fiercely resisting the colonization process, the anthropophagic Caetés were allied with the French who came to smuggle Pau-Brasil (MÉRO, 1974).

Around 1522, the French expedition led by Florentino Joâo Verrazzano reached the Alagoas coast in search of redwood. Fleeing the bodyguard expeditions, they made their way up the River Sâo Francisco, settling in Penedo, in a place far from surveillance, which offered conditions for the easy work of felling and shipping the merchandise (MÉRO, 1974). The author defends the idea that "[...] the Penedo ethnic group was a mixture of French and native Indians" (1974, p.20), and later the introduction of Portuguese and blacks, together with the Indians, formed the basis of the "Penedo ethnic group" (MÉRO, 1974, p.21).

However, the territory of Penedo was only recognized when the system of hereditary captaincies was instituted by the 1st Donatary of the Captaincy of Pernambuco, Duarte Coelho Pereira (BEZERRA, 2006).

It is said that this donatary had received orders from the King of Portugal, Dom Joâo III, to expel the French who were in Itamaracà, Pernambuco. After fulfilling his orders, Duarte Coelho headed for the River Sâo Francisco with the aim of stopping at the end of the captaincy, at the point where Penedo is located. When he crossed the bar of the great river in 1555, he left the foundations of the future town there, as well as numerous families from Portugal who had landed in Pernambuco (VALENTE, 1957).

According to Bezerra (2006), the beginnings of the city's occupation date back to the first years of Portuguese occupation, around 1545, a village known as Sâo Francisco, which was elevated to the category of Vila de Sâo Francisco on April 12, 1636, but by the end of the 16th century it was already called Penedo, due to the large rock on which it sits on the left bank of the river (Image 28).

Image 28: The relief of Rocheira - Penedo.

Source: Archive of the Landscape Study Research Group - UFAL.

The town encompassed all the territory on the banks of the river Sâo Francisco, from where the municipalities of Piaçabuçu, Porto Real do Colégio, Triunfo, Sâo Braz, Traipù, Pâo de Açùcar and Piranhas arose; and even more the central part and the whole of the west, from where the municipalities of Anadia, Limoeiro, Santana do Ipanema, Agua Branca and Paulo Afonso were created (BEZERRA, 2006, p.4).

There are still some historical references to dates, according to Valente (1957), giving Penedo the name of Vila in 1614 and even 1615 in official documents, but, according to the author, the most reliable source is that referred to in Duarte Coelho's "Memórias Diàrias", in a transcription of the excerpt copied from the original, which states that Penedo was elevated to Vila in 1636.

After the death of the donatary in 1554, his son Duarte Coelho de Albuquerque took over the captaincy in 1560, organizing an expedition to fight the native Indians. He then founded a "feitoria", which was an administrative warehouse usually built on the coast of the colony to keep an eye on the natives, exactly on the "penedo" that rises on the bank of the river, with Méro (1974) claiming that this historical fact was proof of the existence of a small population center there.

Situated at the entrance to the river Sâo Francisco, the town served as an important military support point in the region, "being the only route to Bahia (seat of the General Government) and the only viable route for the flow of wealth, gateway to the hinterlands and trading post", according to Bezerra (2006, p. 4). The Dutch arrived in Penedo in 1637 under the command of Mauricio de Nassau, taking over the city and erecting a fort named "Mauricio" in honor of their commander, between the lower and upper parts of the city. If there was any risk of a Portuguese attack across the Sâo Francisco, the privileged location facilitated surveillance (BEZERRA, 2006) (Image 29 and 30).

The town of Sâo Francisco was renamed "Mauricia" and it wasn't long before a revolutionary movement arose in Penedo against the Dutch invaders. Reinforcements were sent from Bahia and, with the help of the people of Penedo, they managed to ambush them until, after much fighting, they had no means of fighting back and had to stay in the fort until they were expelled (MÉRO, 1974).

Image 29: Engraving by Johannes Vingboons- "Mauritius", 1637- View of Penedo and Fort Mauricio.

Source: Orig. Manuscript of the Atlas by J. Vingboons-do AlgemeennRijksarchief, The Hague.CA,1637.

Image 30: Detail of J. Vingboons' engraving from 1647 - First streets of the city and the location of Fort Mauricio de Nassau.

Source: Original manuscript of the Atlas by J. Vingboons from the AlgemeennRijksarchief, The Hague.ca.

It was then, according to Bezerra (2006), that the Portuguese regained possession of Penedo when, in 1645, they expelled the Dutch by destroying the fort and erecting a stone cross at the top of the town. Still in the same century, religious orders (Franciscans, Carmelites and Benedictines) were introduced to the town of Penedo to begin evangelizing and, along with the creation of the houses of each Order, came the advent of the Baroque and the taste for the artistic and cultural (Image 31). It was also in the 17th century that the town was renamed Penedo do Rio Sâo Francisco, and was later simply called Penedo.

Image 31: Franciscan Convent (1908 image).

Source: Archive of the Historical and Geographical Institute of Alagoas (IHGAL).

But it was in the 18th century that the town experienced economic growth, being considered the largest intermediate commercial center between the riverside towns thanks to river navigation, when communication by land was still precarious, receiving products from the country's major centers (Ceara, Bahia and Minas) and transporting them to neighboring towns and even outside the state. In addition, cattle ranching developed in the region, a period that became known as the "Leather Cycle". This period of economic heyday, together with society's interest in culture and art, politics and economics, contributed greatly to Penedo's development in the 19th century (BEZERRA, 2006).

According to the same author, the town of Penedo was elevated to the category of Comarca in mid-1833, at the same time as Alagoas, Atalaia and Maceió, but it was only on April 18, 1842 that it was elevated to the category of city, assuming the position of major economic center of the region. Almost a year after the opening of the São Francisco River to merchant ships from all nations, which took place on September 7, 1867, the Customs House was set up in Penedo on July 6, 1868, motivated by the city's intense commercial activity.

Nineteenth-century Penedo became the scene of a veritable "socio-cultural, political and economic revolution" (MÉRO apud BEZERRA, 2006, p.6).

"Progress was motivated by the industrial era that was taking hold, and the community showed itself to be open to the change from the Baroque philosophy, which had been in force until then, to a mindset dedicated to the technical and the academic" (BEZERRA, 2006, p.6). During the same period, the Lyceu de Penedo was founded, the Telegraph Station was installed, the city's "newspapers" appeared and the Sete de Setembro Theater was inaugurated, among other events, according to Bezerra (2006) (Image 32).

Image 32: Sete de Setembro Theater (Image from 1910).

Source: Archive of the Historical and Geographical Institute of Alagoas (IHGAL).

By the mid-1950s, Penedo already had four textile factories and other food factories, and its industrial production (textiles, oils, rice processing) and land products (cotton, manioc flour, cereals, coconuts, etc.) were transported through its port, which was still the terminus of maritime navigation, to the riverside towns and to the Alagoas sertão. It also serves as the main ferry crossing point on the River Sâo Francisco, linking the state of Alagoas to Sergipe (MÉRO, 1974) (Image 33).

Image 33: Penedo River Port, 1932.

Source: Archive of the Historical and Geographical Institute of Alagoas (IHGAL).

At the end of the 1960s, with progress and modernity, mainly in the area of transport, "the increase and improvement of fast communication routes linking important production and industrial centers to other centers and ports that could provide greater speed", according to Bezerra (2006, p.7).), led to a reversal of the development seen in the city until then, culminating in the 1970s with the construction of the BR-101NE bridge, linking Porto Real do Colégio/AL to Propria/SE and other states (BEZERRA, 2006).

Thus, the Port of Penedo is put on the back burner when it comes to transporting production. The consequent emptying of the port trade, together with the collapse of the textile industry in the country (and the closure of Cia. Industrial Penedense) resulted in the city's economic decline (Image 34). Despite the economic decline witnessed between the 1960s and 1970s, the town survived, and from the end of the 20th century until today, the beginning of the 21st century, it has undergone a new cycle of development, with the installation of sugar and alcohol mills, investment in intensive fish farming and environmental and cultural tourism (BEZERRA, 2006).

Image 34: Cia. Industrial Penedense (1920 image).

Source: Archive of the Historical and Geographical Institute of Alagoas (IHGAL).

2.1.1 Urban expansion and the typological evolution of buildings

The urban center of Penedo is physically made up of two planes: the lower part and the upper part. The lower part is located on the banks of the Sâo Francisco - formed by extensive floodplains made up of lagoons fed by small streams, tributaries of the Sâo Francisco and low, slightly undulating tablelands. The middle/high part of the city - formed by a rocky promontory (penedia) that emerges abruptly from part of the river bank (35m above) and meets a flat plateau formed by a narrow elevated strip of land that widens into a fan in an easterly direction (with heights of over 75m) (BARBIRATO; CAVALCANTE; LINS apud BEZERRA, 2006, p.7).

Penedo's urban layout was established by its topography and geographical features, such as the Catarrinho lagoon, and developed from the port, with the main roads running parallel to the river or radially towards the city's central square (now Praça Jàcome Calheiros). The beginnings of occupation, around 1545, can be seen in the lower part of the city, in the neighborhood known as Rocheira, where there was the port (old anchorage) that established trade in the area and some houses that formed the oldest street in the city, Rua do Sol (BEZERRA, 2006) (Images 35 and 36).

Image 35: Urban expansion and the beginnings of occupation in the city of Penedo.

Source: Image by BEZERRA, Luciane. In: BEZERRA L.,2006, adapted by NORMANDE, Maira.

Image 36: View of the Sâo Francisco River: in the lower part, Rua do Sol in the Rocheira neighborhood, 1920. Source: Archive of the Historical and Geographical Institute of Alagoas (IHGAL).

According to Bezerra (2006), the port of Rocheira ended up silting up due to the various floods of the River Sâo Francisco, preventing boats from squeezing in along Rua da Praia (also known as Rua do Comércio, now Avenida Duque de Caxias), where the city's commerce also moved to at the beginning of the 19th century.

Highlighted on the map of the beginnings of Penedo's occupation, Avenida Floriano Peixoto, the subject of this study, appears in the city's initial occupation as one of the first roads to develop, probably due to its proximity to the old Rua do Comércio, where there was commercial activity (Image 35).

In Rua do Comércio, several colonial-style townhouses were built, some of which functioned as commercial houses on the first floor and residences on the upper floors, as well as single-storey houses facing a square and the river. In 1800 it was forbidden to build next to the river in order to prevent the occupation of this land, where boats and canoes embarked and disembarked, and for the use of the market (Image 37).

Image 37: Colonial townhouses on Rua do Comércio, 1915.

Source: Archive of the Historical and Geographical Institute of Alagoas (IHGAL).

According to Bezerra (2006), the city's initial urban layout underwent almost no changes during its expansion, except for the demolition of Fort "Mauricio" and the alteration of some of the buildings on the plots (adaptation to the new architectural styles and the new rules in force regarding health in housing), "[....] we were even able to define, within the current layout, the streets that originated from the first roads" (BEZERRA, 2006, p.10).).

Then, during the mid-1950s and early 1960s, new neighborhoods sprang up, such as Santa Luzia, the access avenue to the city, Avenida Getúlio Vargas, was opened, and several squares were transformed into squares, such as the current Praça Frei Camillo de Léllis, formerly Largo do Convento (BEZERRA, 2006) (Image 38).

It can be seen that Floriano Peixoto Avenue already appears on the map above, which shows the expansion of neighborhoods in Penedo, forming part of the Historic Center and also of the initial layout of the city, which according to Bezerra (2006) hardly changed during this expansion.

Image 38: Urban expansion: the emergence of new neighborhoods in the city of Penedo.

Source: Image by BEZERRA, Luciane. In: BEZERRA L.,2006, adapted by NORMANDE, Maira.

With regard to the typological evolution of the buildings, Bezerra (2006) states that Penedo began to change its appearance from the 19th century onwards. Until then, the city was marked by the colonial style and the legacy left by the Portuguese colonizers and the French and Dutch invaders, as well as by the presence of the Franciscan missionaries who left their mark on Baroque architecture, in temples and convents, and in buildings constructed from the 16th century onwards (Images 39 and 40).

Image 39: Simple single-storey colonial house.

Source: BEZERRA, Luciane (personal archive), 2006.

Image 40: Church of Sâo Gonçalo Garcia dos Homens Prados - baroque architecture on Avenida Floriano Peixoto.

Source: BEZERRA, Luciane (personal archive), 2006.

The phase of growth caused by economic progress and the ease of communication with the major national and international urban centers provided by the intense river navigation of the port of Penedo, meant that the city assimilated the new architectural styles in force in the major capitals (BEZERRA, 2006). "There was then a notable change in the line of Penedo's civil architecture, with an undeniable neoclassical influence, eliminating the eaves and adding platbands, and houses were also built in the same style"

(BEZERRA, 2006, p.12) (Image 41).

Image 41: Colonial townhouse with later mass elements and platband.
 Source: BEZERRA, Luciane (personal archive), 2006.

"At the end of the same century and the beginning of the 20th century, Eclecticism, Art Nouveau and Art Deco appeared (Images 42 and 43). And in the middle of the 20th century, after experimenting with proto-modernist elements, the first examples of Modernist architecture began to appear" (BEZERRA, 2006, p.12) (Image 44 and 45).

Image 42: House with early 20th century features: Neocolonial and "Art-Noveau" integrated.
 Source: BEZERRA, Luciane (personal archive), 2006.

Image 43: Commercial Association - 2ª half of the 20th century: "Art Deco" and cubist influence represented on Avenida Floriano Peixoto.

Source: BEZERRA, Luciane (personal archive), 2006.

Image 44: House with characteristics from the 2nd ª half of the 20th century: Neocolonial with protomodernist lines.

Source: BEZERRA, Luciane (personal archive), 2006.

The city of Penedo therefore has a wealth of architectural typologies, which tell the story of its urban and architectural evolution and its society, from the Colonial period to Modernism (BEZERRA, 2006). In some cases, this diversity of typologies appears in a single street, as we can see on Avenida Floriano Peixoto with its examples of baroque and art deco architecture, among others (Image 46).

Image 45: Hotel Sâo Francisco - 20th century: Modernism of the late 1950s and early 1960s observed on the avenue under study.

Source: BEZERRA, Luciane (personal archive), 2006.

According to Bezerra (2006), the city of Penedo suffered a slowdown in its progress because it went through a period of economic decadence from the mid-1960s to the 1970s, which ended up putting the brakes on the city's growth to a certain extent and preserving its current appearance until then.

Because of this and the lack of conservation, some parts of the city have degraded and suffered losses, causing gaps in the urban fabric, many of which have been occupied inappropriately (BEZERRA, 2006).

The object of this study, Avenida Floriano Peixoto, is the diversity of styles of buildings that have marked different periods of its historical evolution. Some of the buildings shown in the typological evolution of Penedo's architecture make up the landscape of this avenue, such as the Church of Sâo Gonçalo Garcia dos Homens Pardos, the Commercial Association and the Hotel Sâo Francisco.

This wealth of typologies in a single avenue shows the importance of this road in the historical context of the city and the proposal of this work, which is the requalification of that space through guidelines for the implantation of afforestation and living spaces that once existed (Image 47).

Because it is so significant in the historical context and in the urban and architectural expansion of the city of Penedo, Avenida Floriano Peixoto is now part of the historic center, protected as a cultural heritage site.

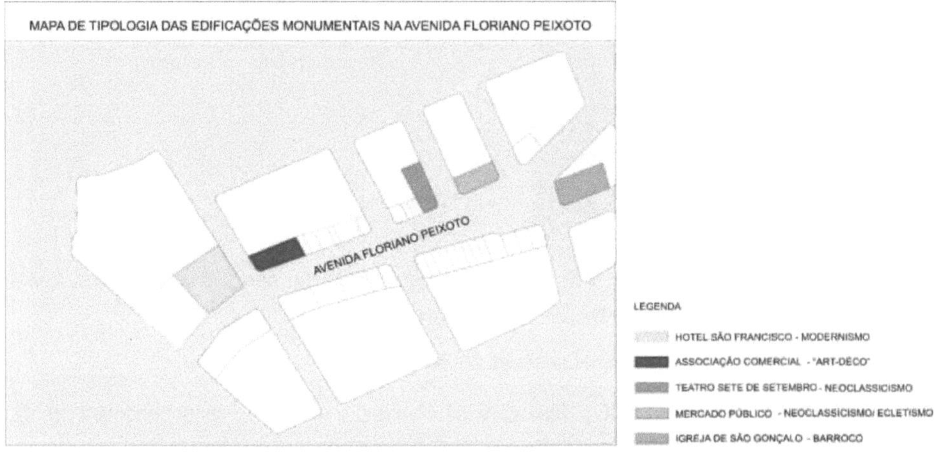

Image 46: Typology of buildings on Avenida Floriano Peixoto.

Source: Image by BEZERRA, Luciane. In: BEZERRA, L., 2006, adapted by NORMANDE, Maira, 2007.

Image 47: Floriano Peixoto Avenue: Buildings of different types. Source: BEZERRA, Luciane (personal archive), 2006.

2.2 THE TOMBED HISTORICAL SITE

Memory is a fundamental engine of creativity: this statement applies to both individuals and peoples who find in their natural and cultural heritage - material and immaterial - the reference points of their identity and the sources of their inspiration (UNESCO apud BESSA, 2004, p.9).

"Throughout history, societies produce knowledge that manifests itself in the arts, politics, religion, behavior and in all everyday relationships" (MACEIÓ, 2004, p.16), in other words, they express their own culture, which represents the identity of each society. Cultural heritage refers to the set of goods, whether tangible or intangible, that have symbolic value for a given society (MACEIÓ, 2004).

The tombamento is an administrative instrument adopted in Brazil with the aim of protecting those assets or sites considered to be relevant to cultural or natural heritage. Although a listed property remains under its original ownership and possession, it cannot be destroyed or de-characterized, thus guaranteeing its preservation. Decoration can be at the federal level, carried out by IPHAN (National Historical and Artistic Heritage Institute), or at the state or municipal level (MACEIÓ, 2004).

"The city of Penedo is listed as a landmark and the delimitation of the site corresponds to the historic center and the core of the city's formation, with the three levels of protection: Federal, State and Municipal" (BEZERRA, 2006, p.16) (See Annex 1) (Image 48).

In addition to the urban ensemble being listed, the site also has individual properties listed at federal and state level. At the Federal level are the Church of Nossa Senhora da Corrente, the Church of Sâo Gonçalo Garcia and the Franciscan Convent of N. Srª . dos Anjos, and at the State level properties such as the Mother Church of N. Sra. do Rosàrio, the Oratory and the Barâo de Penedo House, among others (BEZERRA, 2006) (Image 49 and 50).

Image 48: Listed areas and individual listed buildings in Penedo.

Source: Image by BEZERRA, Luciane. In: BEZERRA, L., 2006, adapted by NORMANDE, Maíra, 2007.

Image 49: Mother Church of Our Lady of the Rosary.

Source: BEZERRA, Luciane (personal archive), 2006.

On the map of the listed areas of the city of Penedo (Image 48), we can see that Avenida Floriano Peixoto is part of both the Federal listed area and the State and Municipal listed areas, as well as having individually listed buildings, such as the Church of Sâo Gonçalo Garcia dos Homens Pardos and the Penedo Public Market.

Image 50: Church of Sâo Gonçalo Garcia dos Homens Pardos on Avenida Floriano Peixoto.
Source: BEZERRA, Luciane (personal archive), 2006.

Penedo's heritage collection is an undeniable historical landmark of the entrepreneurial capacity of our people and reveals, in a peculiar way, their tendency and identity with art and culture. The knowledge and preservation of heritage expresses a positive feeling of self-esteem, stimulates the appreciation of historical roots and mobilizes the community to build collective citizenship (TOLEDO apud MACEIÓ, 2004, p.2).

In 2002, the city of Penedo was accredited to take part in the Monumenta Program, conceived by the Ministry of Culture (MINC) and financed by the Inter-American Development Bank (IDB), with the support of UNESCO. The aim of the program is to preserve priority areas of the country's urban historical and artistic heritage, including public spaces and buildings, in order to guarantee their permanent conservation and the intensification of their use by the population (BEZERRA, 2006). According to Bezerra (2006), the focus of the Program's interventions in cities is the most significant monuments and public places, either because of their historical representativeness or because of their artistic uniqueness. The restoration and subsequent conservation of this ensemble leads to an increase in the value of adjacent properties, generating the economic surplus required for their subsequent conservation.

According to the same author, one of Monumenta's assumptions is that it contributes to the implementation of private projects, since the mobilization of private interests associated with the protection and use of heritage tends to make its preservation permanent.

The Project Area selected by the Monumenta Program in the city of Penedo, according to Bezerra (2006), encompasses the primitive core of the city's development and its immediate surroundings, thus corresponding to the stretch of the Eligible Area with the highest concentration of historical architectural and urban heritage, as well as listed monuments. In the map below, we can see that Floriano Peixoto Avenue is being covered by the Program's actions, as it is part of the city's Historic Listed Center (Image 51).

Image 51: The Monumenta Program's area of operation in the Historic Center of Penedo (in green).

Source: Image by BEZERRA, Luciane. In: BEZERRA, L., 2006, adapted by NORMANDE, Maira, 2007.

The main actions planned for the avenue are: improving its attractiveness, with the restoration of the Church of São Gonçalo Garcia dos Homens Pardos (delivered in 2004) and the Public Market (in the process of being tendered), and improving accessibility, with the restoration of the street itself (in the process of being tendered). In addition to these, there are actions with private participation that involve the restoration of private properties, such as the São Francisco Hotel, also located on Floriano Peixoto. This demonstrates how significant the avenue and its architectural ensemble are in Penedo's listed historic center due to their historical and artistic uniqueness.

In addition, the Traffic Engineering Project carried out by Cidade Digital Consultores Associados Ltda. and presented to Penedo City Hall, which is yet to come into force, provides for the planning of signage and redirection of road flow, due to some urban problems in relation to traffic areas currently found in the city. As this project incorporates the Sitio Histórico area, Avenida Floriano Peixoto will be redirected into a one-way street and will also benefit from a parking ban on some of its crossroads, such as Dâmaso do Monte, Sete de Setembro and Batista Acioly (LIMA; GOMES, 2007).

Despite the various levels of listing of Penedo's historic site and the intervention of the Monumenta Program taking place in the city since 2002, the area is in a slow process of de-characterization and degradation of the landscape, natural and urban cultural heritage (BEZERRA, 2006).

2.2.1 Characterization of the current listed area

The Historic Center of Penedo, as well as the entire municipality, has a water supply from the Sâo Francisco River itself, through the Municipal Treatment System. However, only 12% of the neighborhood has a sewage system (BRASIL apud BEZERRA, 2006). The whole area has overhead electricity and telephone lines. The streets are paved with cobblestones and only one area, located in Praça Barâo de Penedo (Rua Fernando Barros), still has cobblestones (BEZERRA, 2006) (Image 52).

Image 52: Old stone paving in Praça Barâo de Penedo.
Source: BEZERRA, Luciane (personal archive), 2006.

There are no processes of densification or emptying, no verticalization or stagnation, and no generalized degradation. There is a process of modification of the listed urban landscape, caused by the de-characterization of the features of the buildings and the urban ensemble, generated by the addition of sidewalks, elements of signs and billboards, awnings, platbands, antennas, various stone and ceramic coatings, eye-catching paintings, sidewalks of various sizes and coatings, the placement of air conditioners and exhaust fans on the façades, and modifications to the openings and rhythm of the openings, etc. (BEZERRA, 2006, p.19) (Image 53 and 54 and 55).

There are no fortifications or industrial areas in or around the protected area. According to Bezerra (2006), due to the silting up of the River Sâo Francisco and the decrease in river navigation in the region, the city no longer has the structure of the old river port. However, there is still a ferry crossing for pedestrians and vehicles from Penedo to the city of Neópolis, in the state of Sergipe, which still provides considerable activity at the Quay.

Image 53: Use of air conditioners and satellite dishes.
Source: BEZERRA, Luciane (personal archive), 2006.

Image 54: Excessive visual communication. Source: BEZERRA, Luciane (personal archive), 2006.

The main problem found today in Penedo's Listed Centre is the widespread de-characterization of the urban ensemble caused by the city's disorderly growth and, above all, by commercial activity, which changes the appearance of old buildings and historic monuments.

Image 55: Uncharacterized landscape of Floriano Peixoto Avenue. Source: Personal archive, 2007.

Looking at image 55, one can see the transformation of the listed landscape of the area under study, through the de-characterization of the façades of the buildings with the addition of elements such as signs, awnings, poles and overhead electrical wiring, as this is the commercial/services center of the city of Penedo.

Most of the stores, services and institutional facilities are concentrated in the Centro Tombado area, which is consolidated in the area and serves the municipality and neighboring towns. Near the waterfront, commercial and service uses predominate, with a few residences. Towards the interior of the city, there is a predominance of residential use and scattered commercial and institutional houses, which according to Bezerra (2006), is the result of the initial occupation and formation of the city (Image 56).

The map shows that Avenida Floriano Peixoto is predominantly used for commerce and services, making it a busy thoroughfare every day of the week during business hours, and much less so at weekends. There is also religious use with the Church of São Gonçalo Garcia dos Homens Pardos, which moves the avenue a bit at weekends with religious activities and tourist visits, as well as leisure activities with the Sete de Setembro Theater (Image 57 and 58).

Image 56: Land Use Map of Penedo's Historic Center.

Source: Image by BEZERRA, Luciane. In: BEZERRA, L., 2006, adapted by NORMANDE, Maira, 2007.

Image 57: Predominance of commercial/service use on Avenida Floriano Peixoto.

Source: Archive of the Landscape Study Research Group, 2005.

Imagem 58: Uso do solo na Avenida Floriano Peixoto
Fonte: Imagem de BEZERRA, Luciane. In: BEZERRA, L., 2006, adaptada por NORMANDE, Maira, 2007.

Image 58: Land use on Floriano Peixoto Avenue.

Source: Image by BEZERRA, Luciane. In: BEZERRA, L., 2006, adapted by NORMANDE, Maíra, 2007.

According to Bezerra (2006), informal commerce and street markets (Feira do Barro, located on the waterfront, and Feira da Praça Costa e Silva) also exist in the listed area, occupying the streets and sidewalks in a disorderly manner and sometimes without the necessary conditions of cleanliness and hygiene, with stalls and even small buildings installed in areas prohibited by the Municipal Heritage Protection Law of Penedo (Image 59).

The Historic Site is home to the majority of public leisure and religious spaces, such as squares, theaters, museums and churches, which are very popular with the population and visited by tourists. There are also places for training, such as colleges, courses and schools, cultural institutions, newspapers and radio stations, as well as services for tourism, such as hotels, inns, restaurants, travel agencies, etc. (BEZERRA, 2006) (Images 60 and 61).

Image 59: Disorderly occupation of public space in Costa e Silva Square.
Source: BEZERRA, Luciane (personal archive), 2006.

Image 60: Barâo de Penedo Square located within the listed perimeter.
Source: http://www.turismoalagoas.hpg.ig.com.br/penedo-pretos.

There are two petrol stations located on the riverfront within the landmark perimeter, which not only contribute to the de-characterization of the landscape, but also facilitate the heavy traffic of motor vehicles and pose a risk of accidents (explosions, fires) at the heart of the city's formation (BEZERRA, 2006) (Image 62).

Image 61: Pousada Colonial - Penedo, one of the accommodations in the listed center. Source: http://www.turismoalagoas.hpg.ig.com.br/penedo-pretos.

Image 62: Predominance of commercial use and petrol station as an uncharacteristic element of the landscape.

Source: BRASIL, 2002.

Despite the commercial and service activity observed as a strong de-characterizing element of the landscape of Penedo's Historic Centre, the great diversity of uses observed in the Listed Centre demonstrates the dynamism and attractiveness that these activities bring to the place, bringing life to the city.

With regard to the population living in the Historic Center of Penedo, Albuquerque (apud BEZERRA, 2006) states that it is mainly concentrated on the strip bordering the river, towards the interior of the city. This is mainly made up of former residents, many of whom are senior citizens and retirees, who have a good level of education and a reasonable economic income when compared to that prevailing in the rest of the city.

The public transport service satisfactorily serves the entire region and its surroundings, despite the fact that the bus fleet is made up of old vehicles. It is also served by the Municipal Urban Cleaning system on a daily basis, but there is still the accumulation of garbage in an inadequate manner near the formal and informal food businesses, which are possible generators of garbage on the banks of the River Sào Francisco (BEZERRA, 2006).

According to the author, car traffic does not have an orderly and efficient signaling system. The traffic of heavy vehicles, such as trucks and buses, is constant, mainly due to the ferry crossing that takes place at the city's pier, and the supply of local shops located near the waterfront, Avenida Floriano Peixoto, Rua Sabino Romariz and Praça Costa e Silva (Image 63).

Image 63: Ferry crossing on the River Sào Francisco.

Source:

http://www.casadacultura.org.br/AL/penedo/penedo/24_Balsa_RioSàoFrancisco_Penedo_AL.

There are also no suitable, pre-determined parking spaces in these areas. To prevent vehicles using the ferry crossing the entire Historic Center, heavy vehicle traffic can be diverted via the Màrio Freire Leahy highway, but this almost never happens due to ignorance and lack of supervision (BEZERRA, 2006).

We can see that Avenida Floriano Peixoto, the area to be studied in this work, despite having great potential for attractiveness with its commercial and service uses and tourist attractions such as the Sâo Gonçalo Garcia Church and the Sete de Setembro Theater, also functions as a place for heavy vehicles to pass through and park. This should not happen in this listed area, as heavy vehicle traffic damages the structure of these centuries-old buildings, contributing to their degradation (Image 64).

Image 64: Floriano Peixoto Avenue used as a parking lot.

Source: Personal archive, 2007.

The aforementioned Transport and Traffic Project, which will come into force, provides for the reordering of road flow and planning of signage in Penedo's listed center, defining parking and loading and unloading areas. The project that encompasses Avenida Floriano Peixoto demonstrates the problems found there today, with disorderly parking of vehicles, conflicting traffic flows, sidewalks obstructed by street vendors and cross-streets blocked by street markets. This will undoubtedly contribute to the realization of this work, which aims to upgrade that road and provide more space and comfort for pedestrians, currently occupied by vehicles and informal commerce.

With regard to the existing green areas in the Historic Center, Bezerra (2006) highlights the remnants of the traditional backyards and wooded gardens, which mainly contain fruit trees, and the nuclei or isolated specimens in the squares. The other most significant patches of green are the Conventual Fence

70

of the Franciscan ensemble of Nossa Senhora dos Anjos, and the nucleus of the little square located in the lower part of Rocheira, which demonstrates the lack of vegetation in the urban fabric of Penedo's historic center (Image 65).

Image 65: Aerial view of Penedo's historic center, highlighting the green patch of the Conventual Fence of the Franciscan ensemble of Nossa Senhora dos Anjos.

Source: Penedo City Hall Archive, 2002.

2.3 THE OBJECT OF STUDY: FLORIANO PEIXOTO AVENUE

We continued on to Floriano Peixoto, the largest square in the city. Here, under the generous shade of São Gonçalo Garcia, until recently profane festivals came to life with performances by Reisados, Cheganças, Guerreiros, Caboclinhos, Pastoris (SALES, 2003, p.143).

Sales' (2003) reference to the current Avenida Floriano Peixoto as Praça Floriano Peixoto is probably due to the fact that there used to be a Largo de São Gonçalo there, which also gave its name to the avenue. The Largo flowed into the Church of São Gonçalo Garcia dos Homens Pardos, and later, in the middle of the 20th century, it was transformed into a square (Images 66 and 67).

Image 66: The square and church of São Gonçalo at the beginning of the 20th century (1912).
Source: Archive of the Historical and Geographical Institute of Alagoas (IHGAL).

Image 67: Another view of Largo de Sâo Gonçalo (1912).

Source: Archive of the Historical and Geographical Institute of Alagoas (IHGAL).

Quite significant within the urban context of the city of Penedo and, according to Sales (2003, p.139), "[...] the largest historical site in the city", Avenida Floriano Peixoto bears the marks of various architectural styles in its buildings, "[...] especially after the mercantilist force of commerce began to kill its history" (SALES, 2003, p.144).

As we have already seen, the city of Penedo developed from the old port, in a neighborhood known as Rocheira, with the main roads developing parallel to the river or radially towards the city's central square. Trade ended up being established in that town, where a variety of ships docked and where the oldest street in the city was formed.

However, with the silting up of the port of Rocheira, the city's commerce was established in the early 19th century on what is now Avenida Duque de Caxias (formerly Rua do Comércio), as its proximity to the river Sâo Francisco guaranteed natural access for loading and unloading goods.

Therefore, based on this urban expansion of Penedo, it can be deduced that there is a possibility that the current Avenida Floriano Peixoto developed with the implantation of residences and commercial houses that were installed in the area due to its proximity to Rua do Comércio, where this activity already existed. And even when Bezerra (2006) states that the city's main roads developed parallel to the river from the port, we can see the avenue being the first parallel after Rua do Comércio, now Avenida Duque de Caxias, which borders the river (Image 68).

As has already been mentioned, the urban layout of the city did not undergo many changes during its expansion, and it is possible to define within the current layout the streets that originated from the first roads.

In the engraving of the original by J. Vingboons, from 1647, these first roads and streets of Penedo were traced. What should probably be the road that originated the current Avenida Floriano Peixoto is highlighted (yellow) when compared with the map showing the current configuration of the avenue (Image 68). However, this is just an assumption and cannot be proven to be true (Image 69).

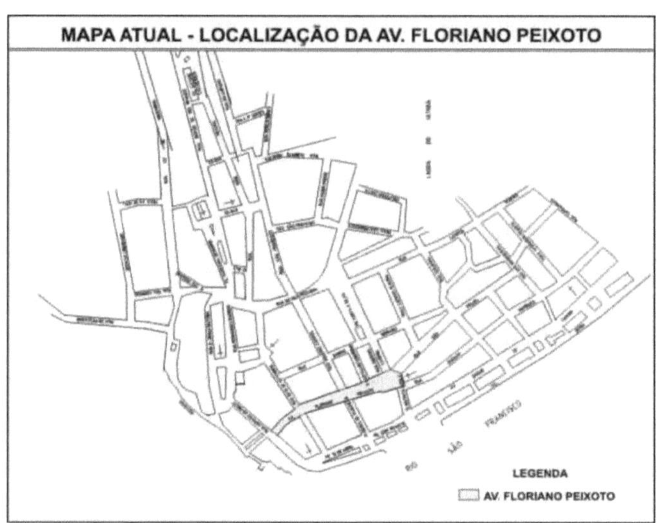

Image 68: Avenida Floriano Peixoto appears as the first parallel to Avenida Duque de Caxias, which borders the Sâo Francisco river.

Source: Monumenta Penedo Program archive, 2002.

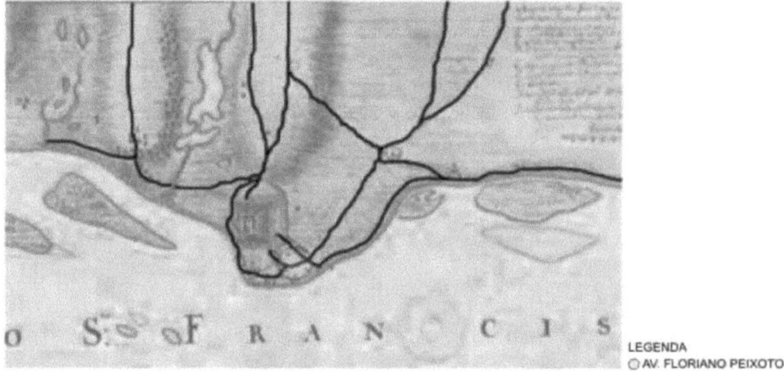

Image 69: Modified detail of the engraving by J. Vingboons from 1647 - showing the first roads and streets of the city of Penedo.

Source: Original manuscript of the Atlas by J. Vingboons from the AlgemeennRijksarchief, The Hague.ca.

The configuration of the old Largo de Sâo Gonçalo has changed a lot over the years. We can see in photographs from the 20th century (some without reference to the exact date) the presence of elements that characterized the Largo, initially the central arborization (1912) (Image 66 and 67), then trees appearing on side avenues, lampposts and a bandstand, where the Penedo music bands played, composing another landscape (Image 70).

Image 70: Arborization of the sidewalks and a bandstand in Largo de Sâo Gonçalo- early 20th century. Source: Archive of the Landscape Study Research Group - UFAL.

Image 71: Garden bed and bandstand in Largo de Sâo Gonçalo (early 20th century). Source: Archive of the Landscape Study Research Group - UFAL.

Later, a landscaped central median can be seen, where it still appears as Largo de Sâo Gonçalo, and at another time the raised sidewalk, flowerbeds distributed along the central area, and the bandstand can be seen (Image 71 and 72).

In other images, the avenue already has a new configuration: the presence of trees on side avenues is again noticeable, and now the presence of the Bust of Floriano Peixoto, located in front of the bandstand, without the flowerbeds in the central area (Image 73).

Image 72: Raised sidewalk and flowerbeds in the Largo (early 20th century). Source: Archive of the Landscape Study Research Group - UFAL.

Image 73: Bust of Floriano Peixoto and absence of central flowerbeds (early 20th century). Source: Archive of the Landscape Study Research Group - UFAL.

In an image from 1920 we can see a new configuration with the presence of central flowerbeds, but still with the same posts and bandstand that appear in images 72 and 73, and in another (with no date reference), we can already see a side street with trees on the public sidewalks, characterizing the former Largo now as Praça Floriano Peixoto (Image 74 and 75).

Image 74: Central flowerbeds and the presence of the bandstand in 1920.

Source: Archive of the Landscape Study Research Group - UFAL.

Image 75: Central flowerbeds and side street in Floriano Peixoto Square (20th century). Source: Archive of the Landscape Study Research Group - UFAL.

And finally, the configuration that most closely resembles today's Avenida Floriano Peixoto with the square transformed into a road, still with the bust and tree-lined public sidewalks, and now with new lampposts characteristic of the 1950s and where you can no longer see the bandstand in the central area (Image 76).

The dynamics of Avenida Floriano Peixoto in the social and economic context of the city of Penedo is probably what led to so many changes in the configuration of this street over a short period of time. The desire to keep up with progress has gradually transformed this landscape and done away with the elements that once characterized it as a square (Image 77).

Image 76: The square transformed into a road and the tree-lined public sidewalks (1950).

Source: Archive of the Landscape Study Research Group - UFAL.

The trees that have always been part of the city's history, as well as providing shade, played a social role when people sat under their canopies to chat and the shoeshine boys served their customers. The force of commercialism replaced the old trees with lampposts and advertising for commercial buildings, putting an end to the social and cultural interaction that once existed in the old square.

This is what can be seen in current photographs of the avenue, which has lost its tree-lined public sidewalks and old streetlights, with the bust of Floriano Peixoto lost amid the informal commerce. Electricity pylons occupy the sidewalks, cars now take the place of people, and what remains of the old monuments are only a few preserved examples, most of which have their façades de-characterized (Image 78).

This shows how de-characterized the landscape of Avenida Floriano Peixoto is and alerts us to the need for an intervention that softens the presence of elements, such as vehicles and electricity pylons, and adds others that recall the historic and cozy image of the avenue with its trees and urban equipment, in a way that preserves the image of the urban ensemble.

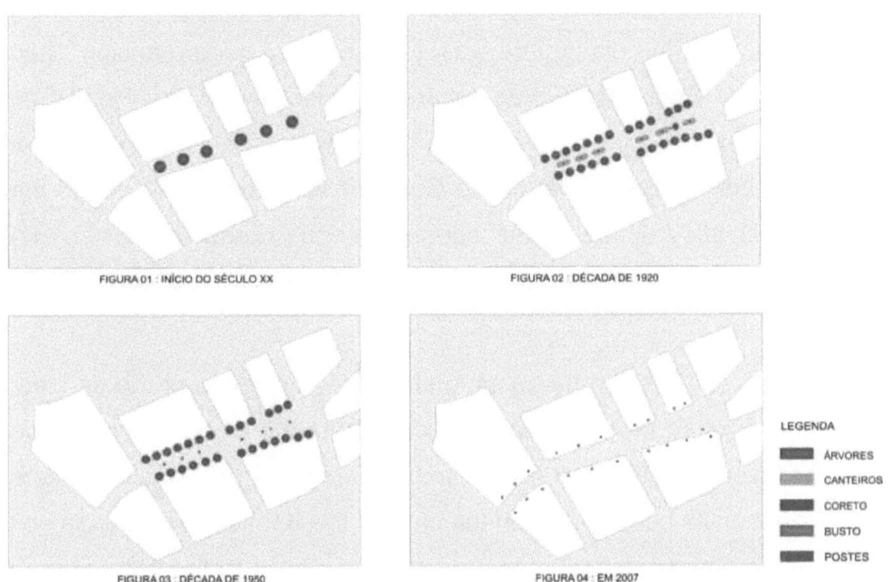

Image 77: Graphic representation of the Avenida's landscape evolution, based on iconographic research.

Source: Image by BEZERRA, Luciane. In: BEZERRA, L., 2006, adapted by NORMANDE, Maira, 2007.

Image 78: The uncharacterized landscape of today's Avenida Floriano Peixoto. Source: Personal archive, 2007.

Although Largo de Sâo Gonçalo still retains the name that was once given to the current Avenida Floriano Peixoto, today it is just a thoroughfare and parking lot. It has lost its trees and other elements that characterized it as a square or even an avenue (SALES, 2003).

Old photos show that the old square was enclosed by a huge mansion. Next to this house was Beco das Flores, where the sidewalk of the Hotel Sâo Francisco runs today. In the void left by the demolition of the mansion, several buildings were built, such as that of the Commercial Association, and in the place of the old alley a street was opened, which contributed to the de-characterization of the colonial style of Avenida Floriano Peixoto (SALES, 2003).

The architectural complex on the avenue shows the influence of the different eras that have marked the historical development of the city of Penedo. The legacy of the colonial style and baroque architecture brought by the colonizers has left its mark on its buildings, as in the case of the Sâo Gonçalo Garcia dos Homens Pardos Church, whose construction dates back to the 18th century (Image 79 and 80).

Image 79: Church of Sâo Gonçalo Garcia dos Homens Pardos (Image from 1915).

Source: Archive of the Historical and Geographical Institute of Alagoa (IHGAL).

According to Sales (2003), the church was originally a chapel built by hermits who called people together for solemnities. In 1758, a brotherhood was formed which began to build it. The high altar is built on a platform and the podium bears the inscription of December 21, 1759, the date on which the foundation stone was laid.

"The church of Sâo Gonçalo Garcia is an example of simple baroque [...] The towers were modified, altering the original volume and balance of the monument" (BORBA; MENEZES apud Sales, 2003, p.141).

Image 80: Church of Sâo Gonçalo Garcia dos Homens Pardos (2006).
Source: Personal archive, 2007.

As already mentioned, the Church of Sâo Gonçalo Garcia was restored by the Monumenta Program in Penedo, with the aim of improving the city's attractiveness, and the work was handed over to the community in 2002.

On Avenida Floriano Peixoto, you can also see the influences brought by the new styles in force in the major capitals, which were assimilated in the city of Penedo at a time marked by economic progress and the ease of communication with the major national and international urban centers, as has already been said, due to the intense movement of the port of Penedo. During this period, the presence of the Neoclassical style is notable, as can be seen in the building of the Sete de Setembro Theater (Image 81).

Image 81: Sete de Setembro Theater (Image from 1910).

Source: Archive of the Historical and Geographical Institute of Alagoas (IHGAL).

The theater was inaugurated on September 7, 1884, and was the first to be built in Alagoas. It was designed by the Italian architect Luis Lucarini, who belonged to the Imperial Philharmonic Society of September 7 (SALES, 2003). "It has an eclectic, elegant neoclassical style, with art nouveau decoration. Its façade preserves statues of Portuguese china representing the Greek muses [...]" (SALES, 2003).

However, according to Sales (2003), the Sete de Setembro Theater lost some of its original features when it was restored by the State Secretariat in partnership with Banco do Brasil (Image 82).

The influence of Eclecticism and Art Deco can also be seen in the late 19th and early 20th centuries in buildings found on the avenue, such as the Public Market and the Commercial Association (Images 83, 84 and 85). And in the middle of the 20th century, the influence of Protomodernist elements and examples of Modern architecture also appeared in the Commercial Association and the Sâo Francisco Hotel, in that order (Images 86, 87 and 88).

Image 82: Sete de Setembro Theater (2005).

Source: Archive of the Landscape Study Research Group - UFAL.

Image 83: Penedo Public Market (Image from 1910). Source: Archive of the Historical and Geographical Institute of Alagoas (IHGAL).

Image 84: Penedo Public Market (2005).

Source: Archive of the Landscape Study Research Group - UFAL.

The Public Market was inaugurated on January 1, 1898, when the Municipality acquired it, in an eclectic neoclassical style, according to Sales (2003).

The project to restore the Public Market is also part of the actions to improve the attractiveness of the Monumenta Program in the city of Penedo, but it is still in the bidding process.

Image 85: The Sete de Setembro Theater and the Public Market on Avenida Floriano Peixoto today.

Source: Archive of the Landscape Study Research Group - UFAL, 2005.

Image 86: Commercial Association - "Art Deco" and Cubist influence. Source: BEZERRA, Luciane (personal archive), 2006.

According to Sales (2003), the Commercial Association building was built to fill part of the void created by the demolition of an old house that closed the then Praça Floriano Peixoto, as mentioned above, with a predominantly square façade, in the "art deco" style of the 1930s (Image 86).

Finally, the Hotel Sâo Francisco, according to Sales (2003, p.153), was "built with modern architectural features from the 50s and 60s of the 20th century, it housed the Cine Sâo Francisco, the stage for a festival that moved the city in the 1970s". Today, the cinema is closed and only the hotel is open (Image 87).

Image 87: Hotel Sâo Francisco - Modernism from the late 1950s and early 1960s.

Source: BEZERRA, Luciane (personal archive), 2006.

Image 88: Commercial Association and Hotel Sâo Francisco on Avenida Floriano Peixoto.
Source: Archive of the Landscape Study Research Group - UFAL (2005).

Hence the peculiar character of the avenue, which has different types of architecture and the influence of various styles that tell the story of its evolution, from examples of the Colonial style to Modernism.

In addition to its historical particularity, the buildings that form part of the urban ensemble of this road make up a perspective that has, at its end, the Church of São Gonçalo as an important focal point, relating in an interesting game of full and empty, vertical and horizontal, and highlighting the Avenue in the context of Penedo's landscape (Image 89).

Image 89: Plots and profiles of Floriano Peixoto Avenue.

Source: Image by BEZERRA, Luciane. In: BEZERRA, L., 2006, adapted by NORMANDE, Maira, 2007.

We can see on the map and profiles of Floriano Peixoto that this uneven road allows a privileged view of the buildings from the highest to the lowest point, with the Hotel Sâo Francisco building "standing out" in the landscape due to its height, right at the top of the avenue. This is also because, a priori, the Church of São Gonçalo represented the highest and most prominent point on the Avenue with its towers, which was later moved to the Hotel, although the Church is still privileged by the perspective that leads to this focal point.

In profile 01 we see the buildings of the Sete de Setembro Theater and the Public Market, privileged by the configuration of the street, as they are located in the stretch where it widens to the focal point of the Church and also by a game of full and empty with the cross streets that isolate and highlight these buildings, and in the case of the Market, because it occupies an entire block.

In profile 02, the horizontality and linearity of one block enhance a discreet verticality and the voids left by the setbacks of plots in the next block, and buildings dialogue with each other with their heights and styles in a play of volumes.

Thus, we realize that the architectural elements that enhance and differentiate the avenue, at the same time, are those that contribute, in part, to the thermal discomfort of the place, due to the massive presence of platbands and implantations without side and/or front setbacks, a situation aggravated by the total absence of afforestation in the place.

The less uncharacteristic façades of profile 01 establish a rhythm between the buildings of proportions and styles that seem to tell the story of the evolution of Avenida Floriano Peixoto, from the colonial sobrado, through the solid "Art Deco" building of the Commercial Association, to the modern style of the Hotel Sâo Francisco.

Returning to Cullen (1983), we can see how the Avenue has the power to exert an emotional impact with its urban ensemble, through the contrasts between full and empty, differences in scale, styles, colors and textures, represented in its landscape with the interesting dialogue exercised by its components.

We can also refer to the concepts of Lynch (1997), when he talks about the power of visual attraction that a set of buildings acquires, enhanced by the elements that surround it, the sequences that lead to it and memories of past experiences. In this way, we perceive the visual richness of Floriano Peixoto with the surprise to be revealed in the sequence of representative buildings that make up its landscape and its historical significance that allows vast associations to be made with its image.

It is also worth highlighting the power of communication that the landscape exerts on the observer, arousing some kind of emotion in them, when Cullen (1983) says that this can be felt when encountering the landscape of a historic site, which reveals in its urban layout and architectural ensemble marks of past eras and is impregnated with memories and meanings.

It can be seen then how this road is an important component of Penedo's landscape, due to the richness of the dialogue between its urban and architectural ensemble and the city's landscape, where the avenue stands out due to the privileged view of its location and proportions, even for those who observe it from the river Sâo Francisco.

However, as has been observed, the historic landscape of Avenida Floriano Peixoto has been de-characterized. The lack of planning and regulation of the legislation that governs this listed area has led to disorderly growth that has generated problems such as the de-characterization of the original features of the buildings, excessive visual communication on them, disorderly parking of vehicles and loading and unloading, and informal commerce occupying the road and sidewalks.

There is also the visual pollution caused by the confusing network of overhead electrical wiring and the noise pollution caused by vehicles, loudspeakers and commercial advertisements. Thermal discomfort has also been mentioned, due to the lack of shading and the absence of places to socialize/rest, among other issues already mentioned.

With regard to street furniture, Avenida Floriano Peixoto has no benches or fixed seats, and not enough garbage cans, public telephones and road signs.

With regard to the accessibility of the road, the sidewalks are of adequate width, according to the minimum width of 1.50m recommended by the ABNT, but they are not in a good state of repair, nor do they have sufficient accessibility ramps in line with the ABNT standards, thus failing to ensure complete mobility for users, especially people with disabilities or reduced mobility. In addition, there are no anti-slip floors or tactile warning for the visually impaired, precisely on a road that is uneven, and there are stretches where the sidewalks are obstructed by informal businesses.

This diagnosis can be verified by applying questionnaires to those who pass through or work on Avenida Floriano Peixoto. This questionnaire was drawn up using the model applied by Lynch (1997), which consisted of capturing the public image of North American cities, and based on the problems currently found in the area through field reconnaissance.

Thus, 10 questionnaires were administered to those who visit, work or use the commercial activity/services offered on Avenida Floriano Peixoto, on weekdays and weekends, in the morning, afternoon and evening, in an attempt to capture the different perceptions of those who frequent it and the distinctions according to the change in time of day.

The questions asked sought to answer issues such as what the Avenue represents and what its importance is for those people, or what purpose brought them there, and the times they most frequented it.

The results of these questionnaires show the historical importance of the urban and architectural ensemble of Avenida Floriano Peixoto for those who experience it, when 50% of the interviewees attribute the significance of the avenue to its historical monuments. Its importance in Penedo's urban dynamics is also perceptible, related to the intense commercial/services activity currently carried out on the site, for 30% of those interviewed, where 45% claim to frequent the site for work, mainly in the morning, with 47% of those interviewed.

Another question aimed to capture the perceptions of space and sensations provoked in different stretches of the avenue, which were divided by the author according to the recognition of a clear distinction between them, in terms of uses, the size of the buildings, the type of implantation on the plot or the usufruct of the local community (Appendix 2).

Thus, in a first stretch, from the beginning to the end of the first blocks (taking the Hotel Sâo Francisco as the beginning of the road), there are only two buildings with service uses: the Hotel Sâo Francisco and the Banco do Nordeste, and only one with commercial use, and one residence. In a second stretch, which began in the following blocks, there was a clear predominance of service and institutional uses, in buildings such as the Commercial Association, the Forum, the Public Prosecutor's Office, the Public Library and the Post Office building. And in a third section, comprising the rest of the blocks and ending at the Sâo Gonçalo Church, there is a predominantly commercial use, except for the religious one represented by the Church, the cultural/leisure one with the Sete de Setembro Theatre, and some services, such as the Lottery House and the Federal Savings Bank.

Then, based on the route taken by the author along these stretches, the different perceptions and sensations caused by aspects such as temperature, noise and visual pollution, safety and olfactory sensations changed along the way. In this way, the interviewees were shown the map and the reason for this division into different sections, as well as current photographs so that they could imagine themselves walking these routes and describe their sensations along them. In this way, it could be seen that the interviewees, in most cases, agreed with what had been perceived by the author.

The results of the perceptions and sensations provoked in the different stretches into which the avenue was divided prove the distinctions between these stretches and between the different times of day observed by the author.

TRACK 01

MORNING: 25% of respondents consider it quiet. Already 22% say they feel cozy, and 22% say the temperature is mild there during that shift.

AFTERNOON: Almost the same as in the morning, with 25% of those interviewed saying that the stretch was quiet and cozy.

NIGHT: 25% say the temperature is mild, and the coziness increases to 24%, with 22% complaining about the lack of security at this time of day, due to the lack of policing.

TRACK 02

MORNING: 19% are bothered by noise pollution, 17% by visual pollution and 15% complain of an unpleasant thermal sensation.

AFTERNOON: 21% of those interviewed say that the stretch is quieter, 15% feel cozier and 15% still complain about the unpleasant thermal sensation.

NIGHT: the temperature is mild, according to 25% of those interviewed, being considered more peaceful for 24% and cozy for 24% of them, and also unsafe at this time for 20% of them.

TRACK 03

MORNING: 25% are bothered by noise pollution, 24% by visual pollution and 22% by unpleasant thermal temperatures.

AFTERNOON: visual discomfort is still felt by 23% of those interviewed, 17% still complain about the unpleasant thermal sensation and noise pollution is decreasing for 14% of them.

NIGHT: 24% say that this stretch becomes quieter, 24% that it becomes cozier, and 23% think that the temperature is milder. In the same way as the first two stretches, 19% of those interviewed considered it unsafe to travel along stretch 03 at night, and only 5% related this to the difficulty of crossing the road due to the heavy flow of vehicles in the morning and afternoon.

According to these results, it can be seen that stretch 03 concentrates most of the problems perceived on the avenue related to the unpleasant thermal temperature in the morning and afternoon, as this is the lowest level stretch, and where there is a greater concentration of vehicles. In terms of noise and visual pollution, it is also the most affected stretch, as commercial use predominates and the facades most affected by commercial advertising and disorderly informal commerce are concentrated here.

It is noticeable that at night time the problems diminish, as the temperature becomes milder and noise and visual pollution are almost non-existent, as there is no more commercial activity and the visual discomfort caused by commercial advertisements and the confusing network of overhead electricity wires is less noticeable.

Questions were also asked about the accessibility of the road, critical visual elements and problems encountered, as well as its greatest qualities, according to those who experience it, thus enabling them to understand what could be improved or should be preserved or privileged in the preparation of the proposal for the requalification of this space.

With regard to the accessibility of the road, 60% of those interviewed said that they find it difficult to get around on foot due to the intense flow of pedestrians and vehicles, the occupation of the sidewalks by informal food stalls, and the structural conditions of the sidewalks, which are in a poor state of repair and inaccessible to people with disabilities or reduced mobility, due to the lack of ramps and the absence of suitable flooring for these people.

When asked which visual elements they considered to be critical on Avenida Floriano Peixoto, 40% of those interviewed believe that informal, disorderly commerce is the main reason for the visual discomfort found there, while 30% believe that parking is the main reason for this. 15% attribute this discomfort to the de-characterization of the façades, and 15% to the overhead electrical wiring network.

The problems encountered today due to the lack of environmental comfort and landscaping in the area could also be seen when 27% of those interviewed complained about the absence of places to socialize, 23% about the lack of shading and another 23% said that visual pollution was the biggest problem encountered on Avenida Floriano Peixoto.

According to the interviewees, the greatest qualities of this avenue are related to the São Gonçalo Garcia Church, due to its historical and architectural importance and its religious activities, for 29% of the interviewees, to the Sete de Setembro Theater, also due to the richness of its architecture and the cultural activities offered at the venue, for 19% of the interviewees, and to the commercial activity developed on the avenue, for another 19%.

In an attempt to recover and research history, people were asked if they knew how Avenida Floriano Peixoto was configured in times gone by, or if they were aware that it had been tree-lined until the mid-20th century, showing old photographs of the area. They were then asked to give their opinion on the possibility of the avenue being tree-lined again, and whether the interviewees believed that this element would provide environmental and landscape comfort for the place, always questioning the reason for their answers. This made it possible to recognize the acceptability of the proposal to be implemented by the people who experience the reality and problems of the area.

Of the people interviewed, 70% said they were aware of the configuration of the old Floriano Peixoto Square and the once tree-lined sidewalks through old photographs of the place and the model of the Pensando Penedo Project (1993), conceived by the Casa do Penedo Foundation and exhibited in the Caixa Econômica Federal building on the avenue. This reinforces the acceptance of the proposal of this work, which aims to resume

elements, such as tree planting and urban facilities, which once characterized that road, approved by 100% of those interviewed. And 100% also demonstrated that they recognize the importance of trees in improving the environmental and landscape comfort of the area.

Lastly, the question was asked about the best way to plant trees on the avenue, and three options were given through photographs: trees on the sidewalks, in landscaped beds, or in pits, separated from pedestrian traffic, or in central beds on the avenue itself.

According to the results, 50% believe that trees should be planted in pits on the sidewalks, providing greater comfort for passers-by and creating places to socialize. 40% believe that trees should be planted in central flowerbeds on the road, and 10% in flowerbeds on the sidewalks, separated from pedestrian traffic. This also reinforces the proposal of this work, which envisages the implementation of a strip of urban furniture and vegetation, leaving an exclusive lane for pedestrian circulation.

Finally, they were asked to give their opinion on the lack of urban facilities, recognizing in fact the shortcomings found in the area for those who use this space.

In addition to the trees, 34% believe that there are not enough garbage cans along the road and 31% complain about the lack of benches to sit on. With regard to accessibility ramps, 14% say they are insufficient and in a poor state of repair, 14% think the road signs are poor and 7% believe there should be more public telephones. It is important to know which urban facilities are considered insufficient by those who travel along the avenue, so that they can be incorporated into the proposal to upgrade this space, which seeks to improve the public sidewalks by installing these facilities, but with vegetation as its main component.

The importance of the diagnosis made by the users and regulars of the Avenue is due to the fact that the requalification of this space aims to provide benefits and an improvement in environmental and landscape comfort, especially for those who experience it, daily or otherwise, as well as for the whole of Penedan society, in this area considered a Cultural Heritage of the Country.

Based on the problems addressed in the questionnaires, we can see the need for this space to be redeveloped by those who use it, in a way that softens the uncharacteristic landscape of the avenue and, above all, seeks to improve the landscape and the comfort of this place, considering its importance in the urban dynamics of the city and in the daily lives of those people.

The guidelines proposed in this work for the implementation of the tree planting and gathering places that once existed on the avenue also envisage organizing the flow of vehicles on the site, delimiting areas for parking, as well as proposing the replacement of the overhead electricity network with an underground one, thus revaluing that historic landscape.

According to the Traffic Project for the city of Penedo, which has not yet come into force, it is planned to change the direction of flow of Avenida Floriano Peixoto to one-way, as well as redirecting the flow and parking of vehicles on some of the roads that cross it. These changes will help in drawing up the guidelines proposed in this work, since the reduction in vehicle traffic in the area will allow more space for pedestrians.

The significance of the avenue is clear, both in the historical and landscape context of Penedo's listed center, but its importance lies mainly within Penedo society itself, not only because of its artistic, cultural and historical value, but also because of its value in the current economic dynamics of the area. The uniqueness of its buildings and of the street itself as part of the city's urban evolution demonstrates the importance of preserving this landscape, which is currently so de-characterized (Image 90).

The avenue, once a square, needs to be given back to pedestrians, to the people who circulate, live and experience it. Even in its current condition as a thoroughfare, it can once again become more welcoming and inviting for those who would like to once again sit in the shade of one of its trees and contemplate the historic and unique landscape of Floriano Peixoto.

Image 90: Floriano Peixoto Avenue as part of Penedo's heritage site.

Source: Archive of the Historical and Geographical Institute of Alagoas (IHGAL) (1962).

3 GUIDELINES FOR THE REDEVELOPMENT OF AVENIDA FLORIANO PEIXOTO

According to the problems found in the area under study, where there is currently a lack of characterization of the historic landscape and the absence of environmental and landscape comfort, guidelines will be drawn up for the implementation of tree planting and places of conviviality, the location of urban furniture, as well as those providing for modifications to the traffic infrastructure and public lighting on Avenida Floriano Peixoto.

Based on a methodology applied by Kevin Lynch (1997), who studies the visual quality of the city through the mental image its inhabitants make of it, based on the components of the environmental image: identity, structure and meaning, a methodological tripod was drawn up for the analysis and apprehension/reading of this area. This tripod relates the components of the image addressed by Lynch to the theories of conservation and restoration, to the technical knowledge of urban afforestation described in chapter 01, and to the research carried out in the field, both the research carried out by the author and the results of the questionnaires.

In order to grasp the <u>identity of </u>the area under study, we used the concepts that deal with the essence and unity of the cultural asset, defended by restoration theorists such as Brandi (1950), where the aim would be to capture the critical elements and visual qualities of this historic landscape and preserve the urban image of the site.

For this apprehension, it was important to analyze the historical evolution of Avenida Floriano Peixoto due to the concept of the dimension of time introduced by Brandi (1950), in which he considers the cultural asset from the time it was made to the time it will undergo the intervention, also highlighted by Delphim (2005) in the basic concepts of conservation theory. It was then possible to observe in the old configuration of the avenue, the presence of trees as an element present until approximately the 1960s, and urban equipment that characterized it as a square, which justifies the proposal of this work that aims to take up these elements that once made up this landscape, but adding a contemporary reading to the area.

It is therefore important that the intervention does not damage the image of the urban ensemble, which is also related to the intrinsic and extrinsic values associated with the property that need to be preserved, as mentioned by Delphim (2005), since the requalification of that space seeks not to break the potential unity of the urban landscape and to preserve the historical, ethnic, symbolic, artistic and even social and economic values associated with the avenue under study.

This is why the relationship between the heights, rhythms, fullness and emptiness and scales of the buildings that make up this landscape was analyzed here, using maps and profiles, to determine the most appropriate way of implanting trees and urban equipment, so as not to damage the image of this urban complex. Consideration was also given to choosing the tree species that would have the least impact on the landscape, preserving the view of the urban complex and not making the vegetation an obstacle, through the characteristics of its size, shape and foliage. In this way, we would also be preserving the intrinsic and extrinsic values associated with the object of study, since the trees and urban equipment are part of the history and symbolism of the site, and perform a social function with the implantation of places of conviviality with shade and urban furniture.

The technical knowledge about the implementation of tree planting on public roads was related to the structure of the environmental image referred to by Lynch (1997), which includes the relationship of space between the object and the observer, analyzing the characteristics of the species suitable for the urban environment and considering aspects such as urban infrastructure, type of occupation, environmental characteristics and compatibility with the width of the road and the electrical wiring network. As well as the author's reading/analysis of the space currently available for planting trees and urban equipment.

To this end, a study was carried out on urban tree planting and its suitability on public roads, and how to make this element of landscape composition and its multiple functions compatible for environmental balance and comfort with the characteristics of the urban environment.

These characteristics were taken into account when choosing the tree species to be planted in a historic landscape, selecting one that does not interfere so much with the façades of the buildings, since, in the case of historic cities, it is imperative that there is total harmony between landscaping and architecture and that it transcends "[....] merely aesthetic issues, but also has a didactic function, the primary objective of which is to transmit to those who contemplate today the interaction that existed between man, the city and nature in the past".

And finally, the <u>meaning</u> that the object assumes for the environmental image of the observer, achieved through popular participation through interviews with those who experience the reality of the Avenue, seeking to capture the public image and diagnosis of the area, and through the reading/appreciation carried out by the author of this work on visits to the site, considering the methodology of Lynch (1997).

We can see in old photographs the beautiful and welcoming landscape that once characterized the avenue as a tree-lined square and its preserved historical monuments. Today we can see a complete lack of trees, the façades of these monuments have been destroyed, there is a profusion of parked vehicles and a mess of electrical wiring.

Today's Floriano Peixoto is the commercial/services center of the city of Penedo, with several stores, galleries, pharmacies, banks, post offices and public bodies occupying the old buildings in a disorderly manner, changing the original appearance of these buildings. In addition, informal trading stalls invade the street and sidewalks, hindering the flow of vehicles and pedestrians, and further contributing to the de-characterization of this landscape, as we have already seen.

Walking along the avenue at the busiest times, usually in the morning and on market days, means living with the strong sun on the sidewalks, the noise pollution from loudspeakers, the visual pollution from the signs and advertisements on the façades, and the confusing network of electrical wiring. The places where the buildings cast their shadows and the staircases of the Sete de Setembro Theater are occupied by passers-by punished by the lack of shade and places to socialize, who find it difficult to circulate along the sidewalks that have been plundered and taken over by shopkeepers.

In the evening, when the avenue is quieter and when you no longer hear the noise of the traffic and don't see the confusion caused by the commercial activity, it becomes a quiet place, where people are afraid to move around and where there are only masses at the Church of St. Gonçalo, the sporadic attractions at the Sete de Setembro Theater, and the movement of two restaurants.

As seen above, the results obtained from the questionnaires revealed the significance of Avenida Floriano Peixoto, attributed to the urban and architectural ensemble, as well as the importance of commercial/service activity in the dynamics of the city of Penedo, for those who live there or pass through. As well as the problem related to the environmental and landscape discomfort of the site and the de-characterization of the landscape of the old Largo de Sâo Gonçalo, remembered for the shade of its trees.

Approval of the proposal to redevelop this space was extremely important, not least because Lynch's methodology takes into account the experience and appropriation of the population, which is essential for this work, since the intervention is intended to bring improvements for those who pass through or live with the current problem of Floriano Peixoto. No less important was the recognition of the benefits brought by the presence of trees on the road, as controllers of the urban microclimate and as an important element of the area's landscape composition.

After analyzing how the avenue and its urban ensemble are part of the historical, urban and architectural evolution in the context of the city's listed center, one can see the importance of this road in the current dynamics and its unique significance in the city's landscape. The very changing character of the old Largo de Sâo Gonçalo, transformed into a square and then a street, shows how Floriano Peixoto has accompanied the transformations and urban dynamics of Penedo, allowing for numerous readings over the years.

You can also see that vegetation was present on the avenue in its various configurations, at least until the 1960s, and that as commercial activity developed in the area, it was replaced by urban equipment and gave way to advertisements on the façades and to vehicles, which increasingly occupy this space previously dedicated to pedestrians.

The guidelines for the area's redevelopment have therefore been divided into: landscaping guidelines, which will guide the implementation of tree planting and the location of urban furniture, and urban infrastructure guidelines, related to changes in the drainage infrastructure, road system and public lighting on Avenida Floriano Peixoto.

For the landscape design guidelines, it was necessary to study the characteristics of the species to be used and technically plan for its implantation in the urban space of the avenue, with its multiple purposes of commerce, services, infrastructure and circulation. As well as the compatibility of the implantation of trees and other urban furniture on the avenue's public sidewalks.

According to an on-site survey, this road is 9.8 m wide at its highest point (the blocks where the Hotel Sâo Francisco and Banco do Nordeste are located). As it follows the gradient, it increases to 12.5 m in the second section (blocks where the Commercial Association and Forum are located), until the final lower section with 30 m, which ends with the Church of São Gonçalo. It was also observed that the sidewalks are approximately 2.5 m wide throughout the street and are somewhat aligned.

As it is planned to redirect the flow of vehicles through the Penedo Transport and Traffic Project, which will turn the avenue into a one-way street, there was the possibility of widening the sidewalks, leaving the road 5.5 m wide in the first section, between 7 and 9 m in the second, reaching 15 m in the final section. So the proposal is to increase the sidewalks on the entire left side of the road to 3 m and widen the sidewalks on the right side, which will be 6.00 m wide, reaching 14.00 m in the final stretch, where it becomes a pedestrian sidewalk and a square in front of the church.

It was decided to widen the sidewalks along the right-hand side due to the excessive sunlight on the left-hand side in the afternoon, during the summer, when the weather is at its hottest, which can be observed on site and highlighted by those interviewed during the questionnaires, who complain of an unpleasant thermal sensation on the stretch at this time of day.

With this in mind, the sidewalks on the left-hand side will be used for passage and circulation, where trees and urban furniture such as bollards, garbage cans and accessibility ramps will be installed. The widened sidewalks along the right-hand side will also function as a place for people to linger, with circular benches around the trees, as there is less sunlight on this stretch, and in front of the church this sidewalk becomes a square.

An exception is made here for the first few blocks (Hotel São Francisco and Banco do Nordeste), where trees will not be planted on the sidewalks, since the ventilation at this higher level eases the discomfort caused by the sun, and where trees have been proposed for the square adjacent to the bank, which is characterized as a place to stay in this area. The proposal also calls for the bust of Floriano Peixoto, currently on the right side of the Church of São Gonçalo, to be located in the same square.

The intention of creating a square in front of the church is to prevent the circulation of vehicles in that area and create a large space dedicated to pedestrians, with the minimum of obstacles to viewing it. It also harks back to the old Largo de Sâo Gonçalo and emphasizes the importance of this building in the landscape of Avenida Floriano Peixoto.

The proposal also foresees the implementation of tree planting in Travessa Carvalho Sobrinho and Travessa Dionizio Campos, located next to the Public Market, which is currently occupied by a street market, which will also function as a place of permanence, with the implementation of urban furniture, and where the market remains, but with standardized stalls.

As seen in section 1.2.1 of this paper, Avenida Floriano Peixoto can be characterized as a wide street, as it is over 7m long, with wide sidewalks (over 2m), and it is recommended that medium-sized species be planted on both sides of the road, considering the proposed guidelines for laying underground electrical wiring, where compatibility with the aerial network will not be necessary.

The decision was therefore made to plant a small species along the entire street, with the aim of minimizing interference with the view of the historic buildings, in places in front of plot setbacks or between buildings. A medium-sized species was chosen for the Banco do Nordeste Square and the side streets to the market, as there is more space available in the square and the side streets, where the trees appear in the central area (Image 91).

The tree species chosen for planting along the avenue is *Senna macranthera*, popularly known as Cassia or Manduirana, a species native to the Brazilian Atlantic Rainforest, occurring from the northeast to the south, with a height of 6 to 8m. It is an extremely ornamental tree, which blooms profusely between December and April and loses all or part of its leaves during the winter. It grows in full sun, especially in high altitude regions. Its rapid growth, small size and sparse crown make it ideal for planting on sidewalks along the road, as it is a species used for tree planting on sidewalks in Brazilian cities (Image 92).

If this species is to be planted on Avenida Floriano Peixoto, proper maintenance and pruning must be carried out, so that the size and shape of the tree is maintained at around 6m in height and the crown is around 3 to 4m, especially in areas where the sidewalk is narrower, so that it does not interfere with the view of the façades.

Image 91: Study of the trees on the profiles of Floriano Peixoto Avenue. Source: Personal archive, 2007.

The medium-sized species chosen for the plaza and crosswalks is *Tabebuia impetiginosa,* popularly known as the purple Ipê. It is also a species native to the Brazilian Atlantic Forest and is widely found in the state of Alagoas. It is a tree that reaches up to 10 m in its adult phase, also with abundant flowering between June and December and semi-caducous leaves. It is a species widely used in urban landscaping because of its beauty and rapid growth (Image 93). It is also important here to carry out pruning in order to control the growth of this species planted on the sides of the Public Market, which will lead to the ideal size for this case, expected to remain between 7 and 8m high.

Image 92: *Senna macranthera* - used to plant trees on sidewalks in Curitiba- PR.

Source:www.casaecia.arq.br/IMAGES/manduirana1.

Image 93: *Tabebuia impetiginosa* - used in urban landscaping for its colorful
flowers.

Source:www.felipex.com.br/imagens/arv_jacaranda.

The widening of the sidewalks was done with a different floor, creating a transition space with the existing sidewalks, in order to make the intervention clear. According to the Penedo Waterfront Project, drawn up by the Monumenta Program, the sidewalks and sidewalks of Avenida Duque de Caxias will receive interlocking flooring. The decision was then made to install the same pavement on Avenida Floriano Peixoto, since this is the first road parallel to Avenida da Orla, in a different color, highlighted by a strip of interlocking pavement, also in a similar color, which will act as a divider between the project and the existing pavement.

The design proposed for this dividing floor was inspired by the shape of the façades on the "tops" of the buildings, whether they were square platbands, triangular pediments, vaulted or pointed church towers, which establish a rhythm between the façades, as seen in the profile of Penedo and Floriano Peixoto itself.

At the same time, this transition strip delimits the free pedestrian circulation strip and the strip intended for the installation of urban furniture and vegetation, located adjacent to the guide, which acts as a separating element between the sidewalk and the traffic lane, providing greater safety and comfort for pedestrians and freeing the lane from interference and obstructions, as well as allowing the soil to permeate wooded areas, increasing the absorption of rainwater (BRASIL, 2004).

As commercial/service use predominates on the avenue, the decision was made to plant the trees at least 30 cm away from the road to avoid damage caused by vehicles, in grassy pits, which allow rainwater to drain, and in the so-called permanence areas, where circular concrete benches were arranged around the trees.

The choice of urban furniture for the avenue was based on those that in some way referred to the site's historic past, but with the use of contemporary materials that were more functional in terms of maintenance. As in the case of the lampposts, with a design that is somewhat reminiscent of the old lampposts in Praça Floriano Peixoto, but at the same time establishes a new language with the area because it is made of galvanized steel (Image 94).

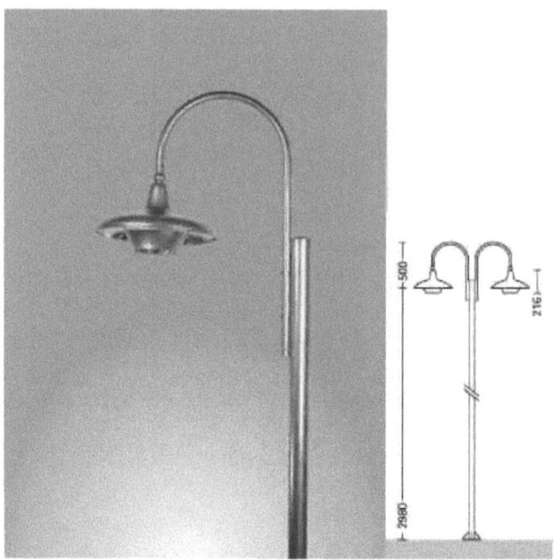

Image 94 : Lumini pole - in galvanized steel, in standard color, 3m high. Source: www.arcoweb.com.br/lightdesign/lightdesign58.asp.

In the same way, the waste bins and public telephones follow the line of the posts, also made of galvanized steel, with a simple, contemporary design. The circular benches around the trees will be made of concrete, as this is a material that is durable and easy to maintain, with a galvanized steel bar that serves as a backrest, thus maintaining a relationship with the material used in the lampposts and garbage cans (Images 95 and 96).

Image 95: lix waste garbage can - galvanized steel body and basket. Source: www.arcoweb.com.br/designmob/.asp.

Image 96: Bank in the city of Ouro Preto (MG), adapted to a circular shape. Source: www.arcoweb.com.br/mobouropreto/.asp.

This equipment was placed within the furniture strip in places where it would not obstruct circulation. In the case of the poles, care was taken to ensure that the façades of the buildings were visible, locating them in places where they would interfere as little as possible and aligning them with the trees, which accompany the perspective of the avenue in side avenues.

Garbage cans and public telephones appear next to benches and crosswalks, which are places where people normally use this equipment and in areas where there is a greater flow of pedestrians.

There was also concern about accessibility for people with disabilities or reduced mobility, with the implementation of ramps on sidewalks, raised crosswalks and tactile floor warnings on suspended obstacles, such as garbage cans and public telephones, in accordance with ABNT standards.

The urban infrastructure guidelines, such as the planning of the flow of vehicles and pedestrians, provide for parking bays in specific places, since the Penedo Transport and Traffic Project does not provide for a ban on parking vehicles on this road. Therefore, places where there is a greater demand for parking were allocated, concentrated mainly in the block where services and institutional use predominates (Forum, Post Office, etc.) and in the Public Market block, in an area where most of the commercial establishments are concentrated.

With regard to the public lighting infrastructure, a proposal was made to install an underground electricity network in the sidewalks, analyzing the most appropriate type for implementation in the area. As can be seen, the existing aerial electricity network on Avenida Floriano Peixoto contributes to the de-characterization of this historic landscape and interferes with the view of the buildings, so the decision was made to install the underground network, which will contribute to the enhancement of this landscape and is reminiscent of the old streetlamps once found on this road.

CONCLUSION

The significance of the avenue is clear, both in the historical and landscape context of Penedo's listed center, and especially within Penedo society itself, not only because of its artistic, cultural and historical value, but also because of its value in the current economic dynamics of the area, which demonstrates the importance of preserving this landscape, which is currently so de-characterized.

The beautiful and welcoming landscape that once characterized Avenida Floriano Peixoto as a tree-lined square and its preserved historical monuments no longer exists. It is important to re-establish tree planting as a key element that has always characterized the avenue and that provides landscape and environmental quality to this listed area.

REFERENCES

ALCIDES, Melissa Mota. **História Naturalis Brasiliae**: A study of the Dutch botanical records of the 17th century. Maceió, 2005, 169p. Dissertation (Postgraduate Program in Development and Environment-PRODEMA). Federal University of Alagoas.

BESSA, Altamiro Sérgio Mol (coor.). **Preserving cultural heritage**: our homes and cities, a legacy for the future! Belo Horizonte - MG: [S.ed], 2004.

BEZERRA, Luciane M. **Centro Histórico de Penedo-Alagoas**: Patrimônio Histórico e Preservaçâo. Penedo: [S.ed], 2006.

BRAZIL. Monumenta. IDB. Ministry of Culture. Penedo City Hall. **Penedo Historical Center Project - Penedo-AL**: Project Profile. Penedo: [S.ed], 2002.

CULLEN, Gordon. **Urban Landscape**. Lisbon: Ediçôes 70, 1983.

COURSE on vegetation applied to landscaping: Arborization of public roads. Curitiba, 1996.

COURSE on urban afforestation. Free University of the Environment. Curitiba, 1995.

D'ASSUMPÇÂO, Livia Romanelli. **Urban Preservation in Diamantina**: theoretical aspects and institutional practice 1938-1970. Salvador, 1995, 266p. Dissertation (Master's Degree in Conservation and Restoration). Federal University of Bahia.

DELPHIM, Carlos Fernando de Moura. **Interventions in Historic Gardens**: **manual**. Brasilia: IPHAN, 2005.

LYNCH, Kevin. **The image of the city**. São Paulo: Martins Fontes, 1997.

LIMA, Josemée Gomes de; GOMES, Roberta Maria Rosas Garcia. **Transportation and Traffic Project - Penedo- AL**. Maceió, 2007.

MACEIÓ. Penedo City Hall. Project Execution Unit - UEP. **Uses, norms and colors**: Historic Center user manual. Maceió, 2004.

MAGALHAES, Manuela Raposo. **Morphology of the landscape**. Lisbon, 1991, 358p. Dissertation (Doctorate in Agronomy). Technical University of Lisbon, Higher Institute of Agronomy.

MANUAL de arborização. Cia Energética de Minas Gerais - CEMIG. Minas Gerais, 1996.

MASCARÓ, Lùcia. **Ambiência Urbana**. 2.ed. Porto Alegre: Masquatro, 2004.

MÈRO, Ernani Otacilio. **História do Penedo**: Elementos de História da Civilizaçâo das Alagoas. Maceió: [S.ed], 1974.

SA, Antônio Fernando de Araujo; BRASIL, Vanessa. **River without History?** Readings on the River Sâo Francisco. Aracaju: FAPESE, 2005.

SALES, Francisco Alberto. **Arruando para o forte**. Recife: Bagaço, 2003.

SIQUEIRA, Vera Beatriz. **Burle Marx**: Spaces of Brazilian Art. Sâo Paulo: Cosac e Naify, 2001.

VALENTE, Aminadab. **Penedo its history**. Maceió: [S.ed], 1957.

WORKS CONSULTED

ABBUD, Benedito. **Creating Landscapes**: A Guide to Working in Landscape Architecture. Sâo Paulo: SENAC Sâo Paulo, 2006.

BRAZIL. Sâo Paulo City Hall. Permanent Accessibility Commission. **Guide to accessible mobility on public roads**. Sâo Paulo, 2004.

CAROATA, José Pròspero Jeovà da Silva. **Chronicle of Penedo**. Maceió: Reediçôes DEC - Revista do Instituto Histórico e Geogràfico de Alagoas, 1962.

DEL RIO, Vicente. **Introduction to urban design in the planning process**. Sâo Paulo: Pini, 1990.

MACEDO, Silvio Soares. **Landscaping in Brazil**. Sâo Paulo: Quapa, 1999.

MONUMENTA. IDB. Ministry of Culture. Penedo City Hall. **Socio-environmental Assessment Report**. Penedo: [S.ed], [2002].

RECIFE CITY HALL. Urban Planning Department. Environmental Development Department. **Arborization of Recife**: Technical Notes for Adjustments in Execution and Maintenance. Recife: [S.ed], [1996].

PENEDO CITY HALL. Participatory Master Plan.

Penedo Participatory Master Plan Technical Reading Report: Mobility Theme. Penedo: [S.ed], [2007].

SANTOS, Milton. **Metamorphoses of Inhabited Space**. Sâo Paulo: Hucitec, 1991.

ANNEX

ANNEX I

Tomb Legislation

The legislation that governs the protection of the historical, cultural and landscape heritage of the city of Penedo, as well as the permitted uses and regulations are:

Federal Constitution of 1988.

Art. 23. It is the common competence of the Union, the States, the Federal District and the Municipalities:

III - To protect documents, works and other goods of historical, artistic and cultural value, as well as monuments, natural landscapes and archaeological sites;

V - Urban complexes and sites of historical, landscape, artistic, archaeological, paleontological, ecological and scientific value.

Decree-Law No. 25 of November 30, 1937.

Art. 17: Under no circumstances may listed items be destroyed, demolished or mutilated, or, without prior special authorization from the Heritage Department

National Historic and Artistic12, be repaired, painted or restored, under penalty of a fine

50% of the damage caused.

Federal Listing - MINC Ordinance No. 169 of December 18, 1995.

THE MINISTER OF STATE FOR CULTURE, using the powers conferred on him by Law No.

6.292, of December 15, 1975, and in view of the manifestation of the Council

Consultivo do Patrimònio Cultural at its 7ª Meeting held on December 7, 1994, resolves:

I - To approve, for the purposes of Decree-Law No. 25 of November 30, 1937, the

Listing of the Historic and Landscape Complex of the City of Penedo, State of

Alagoas, according to the perimeter delimited on page 237 of Case No. 1201-T-86.

The "Conjunto Histórico e Paisagistico da Cidade de Penedo" (Historical and Landscape Complex of the City of Penedo) was listed by IPHAN -

Instituto do Patrimònio Histórico e Artistico Nacional, under No. 541 of the Historical Book, volume

2, folios 26/29, on 30/10/1996, and under No. 113 of the Archaeological, Ethnographic and

Paisagistico, volume 1, pages 77/80, on 30/10/1996, in case no. 1.201-T-86.

The perimeter area (see Location Sheet of the Listed Area) is approximately,

27 hectares, it is estimated that it has 800 homes, corresponding to around 4,000 inhabitants

State and Municipal Listing

Penedo has been granted State Listing of its "Historical, Artistic and Cultural Heritage".

Natural" through Decree No. 25.595 of 08/03/1986. After almost 03 years, it obtained the

Municipal listing - Law no. 939 of November 9, 1989 - which approved

The same content as the state landmark decree, including keeping the same polygon area (See Landmark Area Location Sheet)14.

Municipal Listing Law - Law No. 939 of November 9, 1989.

THE CITY COUNCIL OF PENEDO DECREES:

Art. 1 - In compliance with Resolution 03 of February 25, 1986, of the State

Council for Culture and the Preservation of Historical, Artistic and Natural Heritage, ratified by

State Decree No. 25,595 of March 8, 1986, declared the city of Penedo a listed city.

Art. 2 - The listing referred to in Art. 1 covers the area contained within the boundaries of a polygon (Poligono de Tombamento de Penedo - PTP) in Penedo, made up of three distinct zones:

I - Environmental Preservation Zone - ZPA, are the areas surrounding the Environmental Preservation Zone that ensure environmental protection;

II - Strict Preservation Zone - ZPR, where the Historic Center itself is located;

III - Landscape Preservation Zone - ZPP, are areas of flooded mangroves, woods and other forms of vegetation on hillsides and stretches of steep slopes.

Art. 4 - Conservation, repair and restoration work in a Strict Preservation Zone (ZPR) shall respect the volume and appearance of the property in its own right, and in relation to the scale and shape of the ensemble in which it is located, where appropriate, maintaining the originals:

I - The height and number of floors of the existing building, in the case of repair and restoration work;

II - The construction of the building on the land;

III - The shape and slope of the deck;

IV I - Wall and roof covering materials;

V V - Wall and roof cladding materials;

VI - Circulation, ventilation and lighting.

ANNEX II

DIVISÃO DOS TRECHOS

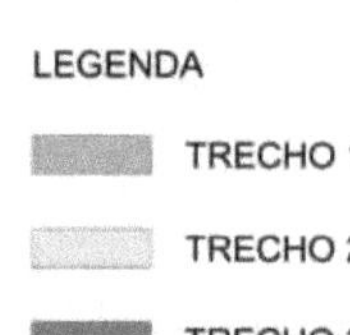

LEGENDA

- TRECHO 1
- TRECHO 2
- TRECHO 3

yes
I want morebooks!

Buy your books fast and straightforward online - at one of world's fastest growing online book stores! Environmentally sound due to Print-on-Demand technologies.

Buy your books online at
www.morebooks.shop

Kaufen Sie Ihre Bücher schnell und unkompliziert online – auf einer der am schnellsten wachsenden Buchhandelsplattformen weltweit! Dank Print-On-Demand umwelt- und ressourcenschonend produziert.

Bücher schneller online kaufen
www.morebooks.shop

info@omniscriptum.com
www.omniscriptum.com

MIX
Papier aus verantwortungsvollen Quellen
Paper from responsible sources
FSC® C105338

FSC
www.fsc.org

Printed by Books on Demand GmbH, Norderstedt / Germany